THE STARSHIP CHALLENGE

How SpaceX's Audacious Tests Are
Changing the Way We Think About Space

William A. Sanders

Copyright © 2024 by William A. Sanders.

All rights reserved. No part of this publication may be reproduced, distributed, or transmitted in any form or by any means, including photocopying, recording, or other electronic or mechanical methods, without the prior written permission of the author, except in the case of brief quotations embodied in critical reviews and certain other non-commercial uses permitted by copyright law.

Disclaimer:

The advice and strategies contained herein may only be suitable for some situations. This work is sold with the understanding that the author and publisher are not engaged in rendering professional services. If professional assistance is required, the services of a competent professional should be sought. The author and publisher specifically disclaim any liability incurred from the use or application of the contents of this book.

Table of Contents

Introduction 8
Why Starship Matters 8
PART 1 13
FOUNDATIONS OF SPACEFLIGHT 13
Chapter 1 14
A Brief History of Space Exploration 14
    The Dawn of the Space Age 15
    Apollo: Humanity's Giant Leap 16
    The Shuttle Era: A New Vision for Space 17
    The Rise of Private Space Companies 18
    A New Space Economy 19
    From Government-Led Missions to Private Innovation .. 20
    Looking Ahead 21
Chapter 2 24
The Basics of Rocket Science 24
    The Challenge of Gravity 25
    Understanding Orbital Mechanics 25
    Propulsion Systems: The Engines of Exploration 26
    Stages of a Rocket's Journey 28
    The Physics of Re-Entry 29
    The Importance of Reusability 30
    The Road Ahead 31
Chapter 3 32
The SpaceX Revolution 32

    The Birth of SpaceX..33
    The Falcon 1: A Turning Point............................... 34
    The Falcon 9: Reusability Redefined.....................35
    The Dragon Capsule: Opening New Horizons..........36
    Starship: The Vision for a Multi-Planetary Future......37
    The SpaceX Model: Disruption Through Innovation..38
    Global Competition and the New Space Race.......... 39
    Legacy and Future of SpaceX.................................. 40

PART 2..**41**
INSIDE STARSHIP'S DEVELOPMENT.....................**41**
Chapter 4..**42**
Building the Unbuildable..**42**
    The Challenge of Reusability...................................43
    The Raptor Engine: Powering the Future................. 44
    Stainless Steel: A Material Revolution......................45
    Heat Shields: Surviving Re-Entry............................. 47
    Balancing Cost-Efficiency with Innovation................48
    Testing and Iteration: The Path to Progress............. 49
    The Vision Behind Starship...................................... 50
    The Unbuildable, Built...51

Chapter 5..**52**
Trial by Fire.. **52**
    Early Prototypes: From Starhopper to SN Series...... 53
    High-Altitude Flight Tests: Learning Through Failure.54
    The Evolution of the Heat Shield............................. 55
    Key Breakthroughs and Milestones..........................57
    The Philosophy of Iteration.......................................57
    Lessons Learned and Future Goals......................... 58
    Triumph Through Trial.. 59

Chapter 6..62
Iteration and Innovation....................................62
    The Philosophy of Rapid Prototyping....................... 63
    The Role of Failure in Accelerating Progress........... 64
    Unexpected Outcomes Shaping Starship's Design... 65
    Scaling Innovation: From Prototype to Production.... 68
    Cultural Innovation: A Mindset for the Future............ 69
    The Broader Impact of Iteration and Innovation........ 70
    The Power of Iteration.. 70

Chapter 7..72
Hot Staging and Re-Entry..................................72
    The Science of Re-Entry: Facing the Heat............... 73
    Starship's Thermal Protection System......................74
    The Aerodynamics of Re-Entry: The Belly-Flop Maneuver..75
    The Innovation of Hot Staging.................................. 76
    Starlink-Enabled Telemetry: Real-Time Insights........77
    Preparing for Mars: The Ultimate Test...................... 78
    The Future of Re-Entry Technology..........................79
    Pioneering the Path to Mars..................................... 80

PART 3..81
THE HUMAN ELEMENT......................................81
Chapter 8..82
The People Behind Starship..............................82
    Elon Musk: The Architect of the Dream....................83
    Gwynne Shotwell: The Steady Hand......................... 84
    The Engineers and Scientists Behind Starship......... 85
    The Culture at SpaceX: Collaboration and Resilience.. 87
    The Impact of Elon Musk's Vision.............................88

The Future of Starship's Team..................................89
The Human Factor in Space Exploration..................90

Chapter 9..**92**
The Public's Spacecraft..**92**
The Power of Inspiration: A Global Dream................ 93
Live Broadcasts: Sharing the Journey...................... 94
Social Media: Building a Community......................... 95
Transparency and Engagement: Redefining Public Relations..96
Inspiring a New Generation....................................... 97
The Role of Failure: A Shared Journey..................... 98
A Vision for the Future... 99
A Spacecraft for All.. 100

Chapter 10..**102**
Ethical and Philosophical Questions...................**102**
Who Owns Space? The Legal and Ethical Implications 103
The Environmental Impact of Space Travel............105
Philosophical Reflections: Becoming a Multi-Planetary Species...107
Space Exploration as a Reflection of Humanity......109
The Questions That Define Us................................110

PART 4...**111**
BROADER IMPLICATIONS......................................**111**
Chapter 11..**112**
Economics of the Final Frontier............................**112**
The Financial Model of SpaceX: Revolutionizing Cost Efficiency...113
The Economics of Interplanetary Colonization........ 115
Commercial Space Ventures: Opportunities and

Challenges.. 117
The Long-Term Vision: A Spacefaring Economy..... 119
The Economics of Exploration................................ 120

Chapter 12.. 122
SpaceX and the New Space Race................................... 122
The Rise of SpaceX: A Catalyst for Change............ 123
International Efforts in the New Space Race........... 124
Geopolitical Implications of a Privatized Space Race.... 127
The Role of SpaceX in the New Space Race........... 129
A New Era of Exploration.. 130

Chapter 13.. 132
Preparing for Mars... 132
Why Mars? The Case for Colonization.................... 133
The Technical Challenges of Reaching Mars.......... 134
The Logistical Challenges of Colonizing Mars........ 136
The Human Challenges of Mars Colonization......... 138
Starship: The Cornerstone of Mars Exploration...... 139
The Path Forward: Preparing for Humanity's Most Ambitious Journey.. 141

Chapter 14.. 144
Beyond Mars... 144
Asteroid Mining: The Next Frontier of Resource Extraction.. 145
Colonizing the Outer Solar System.......................... 147
Interstellar Travel: Humanity's Ultimate Challenge.. 148
Humanity's Role in the Cosmos............................... 150
The Role of Starship in Humanity's Cosmic Future. 151
A Journey Without End.. 152

Conclusion... 154

The Starship Legacy 154
    A Transformative Vision... 155
    Lessons from Starship: Innovation and Resilience .. 156
    A Broader Impact: Inspiring Future Innovations 158
    The Importance of Global Collaboration 159
    A Call to Action: Curiosity, Ambition, and Unity 160
    The Starship Legacy 162

Appendix 164
    Appendix A: Technical Specifications of Starship 164
    Appendix B: Key Milestones in the Starship Program ... 165
    Appendix C: Glossary of Terms 166
    Appendix D: SpaceX's Key Contributions to Space Exploration 166
    Appendix E: Timeline of Space Exploration 167
    Appendix F: Recommended Resources 168
    Appendix G: Ethical Considerations in Space Exploration 169
    Appendix H: Frequently Asked Questions About Starship 169
    Appendix I: The Future of Space Exploration 170

About the Author 172

Introduction

Why Starship Matters

The world held its breath. On a remote stretch of Texas coastline, against the stark backdrop of steel and sand, a silver tower of impossibility stood poised to challenge the limits of human ingenuity. The countdown began, each second resonating not just with the engineers crowded in control rooms, but with dreamers across the globe who dared to believe in something larger than themselves.

At T-minus zero, the engines roared to life, unleashing a maelstrom of fire and thunder. The ground shook, the air crackled, and in a moment of raw defiance against gravity's relentless pull, *SpaceX's Starship soared*. This was no ordinary launch—it was a symphony of ambition, a testament to humanity's refusal to remain tethered to a single planet.

Yet, minutes later, the stakes became clear. The spacecraft, now miles above Earth, faced its ultimate trial: re-entry. Hurtling through the atmosphere at breakneck speeds, the Starship glowed like a falling star, a fiery streak across the heavens. Would it survive the searing heat of re-entry, or would it succumb to the very forces it sought to conquer? The plasma that enveloped it wasn't just a test of technology; it was a metaphor for the crucible of progress itself. Success or failure here wasn't just SpaceX's burden—it was humanity's.

The Bold Question
The Starship program isn't just about rockets. It's not just another chapter in the story of space exploration. It's a bold answer to a question that has haunted humankind for centuries: *What lies beyond the horizon—and do we have the courage to go there?*

For Elon Musk and his team at SpaceX, that horizon isn't just the Moon or Mars. It's a future where humanity transcends its earthly confines, spreading life to other planets, becoming a multi-planetary species. Starship is the vessel for this vision—ambitious, audacious, and wholly unprecedented. It is both a technical marvel and a philosophical statement: *Why should humanity remain bound by the chains of one world when the stars beckon us forward?*

The stakes couldn't be higher. Starship isn't just a spacecraft; it's a proving ground for ideas, technologies, and resilience. It's the key to unlocking a new era in space exploration—one where the impossible becomes routine, and where every citizen of Earth can dream of touching the heavens.

The Stakes: Beyond the Sky

Imagine a future where Starship isn't just a prototype tested in the remote deserts of Texas but the backbone of humanity's greatest leap forward. Picture fleets of Starships ferrying people, cargo, and hope to the red sands of Mars. Envision a time when resources mined from asteroids reshape Earth's economy, and off-world colonies become self-sustaining cradles of human civilization.

But getting there requires solving problems that no one has ever faced. How do you make a spacecraft not just reusable, but reliable enough to fly hundreds of missions without costly overhauls? How do you keep astronauts safe during re-entry, when the friction of Earth's atmosphere can vaporize steel? And how do you convince a skeptical world that such a future is worth the enormous risks and costs?

Starship is the embodiment of these questions. Every test flight, every explosion, and every incremental success brings us closer to answers. The story of Starship is a story of persistence, failure, and triumph on a scale humanity has rarely attempted. It's a story that's still being written—and one that will define the 21st century.

A Journey Through Starship
This book is your guide to understanding why Starship matters—not just to engineers or scientists, but to everyone. It's a deep dive into the heart of SpaceX's most ambitious project, exploring the technical, human, and philosophical dimensions of a spacecraft that's changing the way we think about exploration.

We'll start by looking back at how we got here, tracing the evolution of space exploration from the first tentative steps into orbit to the groundbreaking innovations of SpaceX. From there, we'll dissect the engineering marvels that make Starship possible, from its Raptor engines to its innovative heat shield. We'll examine the culture of SpaceX, where failure isn't just tolerated but celebrated as a stepping stone to progress.

But this isn't just a technical story. It's a human story—about the people who dream, build, and test Starship, and about the billions of us who watch in awe as it leaps toward the heavens. It's a story about the philosophical implications of leaving Earth: What does it mean to become a multi-planetary species? Who decides who gets to go? And what responsibilities do we have to the planets we'll one day call home?

The Legacy of Starship
As you turn the pages, you'll discover more than just the science and engineering of a spacecraft. You'll uncover

the broader significance of Starship as a symbol of human ambition. It represents not just the possibility of traveling to Mars, but the spirit of exploration that has defined our species since the first humans set foot on uncharted land.

Starship isn't just SpaceX's greatest challenge—it's humanity's. And while the road ahead is uncertain, one thing is clear: the journey is worth it. As the engines roar and the plasma glows, Starship reminds us of the one truth that has always driven us forward: the stars are not as far as they seem.

So buckle up. The countdown has already begun. Let's explore the Starship challenge together.

PART 1

FOUNDATIONS OF SPACEFLIGHT

Chapter 1

A Brief History of Space Exploration

From the moment humans first gazed at the night sky, the stars inspired a sense of wonder, mystery, and endless possibility. For centuries, they were regarded as untouchable, a celestial tapestry beyond the grasp of human ingenuity. The ancients wove myths around them, using their light to guide ships, track time, and define seasons. But the dream of traversing the void between Earth and the heavens remained confined to the realms of imagination. Then, in the 20th century, that dream began to materialize. It was an audacious transformation, driven by scientific discovery, political ambition, and an unrelenting desire to reach beyond the known. This chapter chronicles humanity's journey into space, from its humble beginnings to the revolutionary changes that have brought us to the brink of interplanetary exploration.

The Dawn of the Space Age

On October 4, 1957, a faint beeping sound echoed across the world, carried by the simplest of radio signals. It came from Sputnik, the first artificial satellite, launched by the Soviet Union. Weighing just 184 pounds, the shiny, spherical satellite orbited Earth at a speed of 18,000 miles per hour, completing a circuit every 96 minutes. To the Soviet scientists who built it, Sputnik was a triumph of engineering; to the rest of the world, it was a wake-up call. Space, once the domain of gods and legends, had suddenly become reachable.

The significance of Sputnik extended beyond its technical achievement. It marked the start of the space race, a geopolitical contest that would define much of the Cold War. The Soviet Union's success ignited a sense of urgency in the United States, leading to the creation of NASA in 1958 and the launch of America's own satellite, Explorer 1, in January of that year. These early missions set the stage for an era of rapid progress in space exploration, driven by national pride and the desire for technological supremacy.

The space race wasn't just about hardware; it was a battle for hearts and minds. The Soviet Union continued to dominate early milestones, sending the first living creature (Laika the dog) into orbit in 1957 and achieving the first human spaceflight with Yuri Gagarin's Vostok 1 mission in 1961. Each success was a testament to the ingenuity and determination of the scientists and engineers behind them, inspiring awe and competition across the globe.

Apollo: Humanity's Giant Leap

The early Soviet victories galvanized the United States to aim higher. In 1961, President John F. Kennedy declared that America would land a man on the Moon and return him safely to Earth before the end of the decade. This bold challenge set the Apollo program into motion, an unprecedented effort that would define NASA's legacy and establish the United States as a leader in space exploration.

The Apollo program was a monumental undertaking. More than 400,000 engineers, technicians, and scientists worked tirelessly to solve problems that had never been encountered before. How could a spacecraft be designed to survive the intense heat of re-entry after traveling at 25,000 miles per hour? How could astronauts safely navigate the 240,000-mile journey to the Moon and back? These questions pushed the boundaries of technology, leading to innovations in computing, materials science, and communications.

On July 20, 1969, Apollo 11 achieved what many thought was impossible. As Neil Armstrong stepped onto the lunar surface and declared, "That's one small step for man, one giant leap for mankind," the world held its breath. The Moon landing wasn't just a technological triumph; it was a moment of collective human achievement. Over the next three years, six more Apollo missions followed, furthering scientific knowledge and solidifying the Moon as a symbol of human ingenuity.

The Shuttle Era: A New Vision for Space

As the Apollo program concluded, NASA faced a new challenge: how to make space exploration sustainable. The answer came in the form of the Space Shuttle, a reusable spacecraft designed to reduce the cost of reaching orbit. First launched in 1981, the Shuttle was a remarkable engineering achievement, capable of carrying both crew and cargo into space and returning safely to Earth.

The Shuttle era introduced a new paradigm for space exploration. No longer focused solely on bold feats like Moon landings, NASA turned its attention to building infrastructure in space. The Shuttle was instrumental in deploying satellites, conducting scientific experiments, and constructing the International Space Station (ISS). It also facilitated international collaboration, with astronauts from around the world participating in missions.

Despite its successes, the Shuttle program faced significant challenges. The Challenger disaster in 1986 and the Columbia disaster in 2003 were stark reminders of the risks involved in human spaceflight. These tragedies reshaped NASA's approach to safety and underscored the need for continued innovation. The Shuttle program, which concluded in 2011, left a legacy of scientific progress and laid the groundwork for the next phase of space exploration.

The Rise of Private Space Companies

The end of the Shuttle program marked a turning point in space exploration. As NASA scaled back its human spaceflight capabilities, private companies began to step into the spotlight. Leading this charge was SpaceX, founded by Elon Musk in 2002 with the ambitious goal of making life multi-planetary. Musk's vision was clear: to create a new era of space travel that was accessible, sustainable, and geared toward interplanetary exploration.

SpaceX's early years were fraught with challenges. Its first three rocket launches ended in failure, pushing the company to the brink of bankruptcy. But in 2008, the successful launch of the Falcon 1 rocket demonstrated that a privately funded spacecraft could reach orbit. This achievement marked the beginning of a revolution in spaceflight. SpaceX's focus on reusability, exemplified by the Falcon 9 rocket's ability to land vertically and be reused, drastically reduced the cost of space access.

Other companies soon followed suit. Blue Origin, founded by Amazon's Jeff Bezos, focused on developing reusable suborbital and orbital launch vehicles. Rocket Lab, Virgin Galactic, and other private firms brought fresh perspectives and innovations to the industry. Together, these companies transformed space exploration from a government-dominated field into a vibrant commercial ecosystem.

A New Space Economy

The rise of private space companies has reshaped the economics of space exploration. What was once the exclusive domain of superpowers has become a global industry, driven by competition, innovation, and opportunity. Private firms now launch satellites for telecommunications, weather monitoring, and Earth observation, while space tourism is beginning to take off with suborbital flights for paying customers.

This new space economy is also fostering international collaboration. The ISS, for example, relies on private companies like SpaceX to deliver supplies and transport astronauts. Similarly, NASA's Artemis program, which aims to return humans to the Moon, partners with private firms to develop lunar landers and support systems. These collaborations blend public ambition with private innovation, accelerating progress and expanding the possibilities of space exploration.

Looking ahead, the commercial space sector shows no signs of slowing down. Plans for asteroid mining, lunar bases, and Mars colonization are moving from concept to reality, driven by the ingenuity of private companies and the vision of a multi-planetary future.

From Government-Led Missions to Private Innovation

The transition from government-led missions to private-sector innovation represents a fundamental shift in the landscape of space exploration. During the 20th century, national space agencies like NASA and the Soviet space program drove progress, motivated by geopolitical rivalry and scientific curiosity. But the high costs and inherent risks of space exploration made it difficult to sustain these efforts over the long term.

Private companies have introduced a new model, characterized by agility, efficiency, and an entrepreneurial spirit. By focusing on cost reduction and rapid iteration, firms like SpaceX have demonstrated that space exploration can be both ambitious and economically viable. This shift has created a symbiotic relationship between public and private entities, with governments setting the agenda for deep-space exploration while private companies provide the tools and capabilities to achieve it.

This collaboration has unlocked new opportunities, from robotic missions to Mars to the development of space-based industries. It has also sparked a renewed sense of excitement and possibility, inspiring a new generation of scientists, engineers, and explorers to look to the stars.

Looking Ahead

The history of space exploration is a story of relentless ambition, resilience, and innovation. From the beeping signal of Sputnik to the reusable rockets of SpaceX, each milestone has pushed humanity closer to the stars. As we stand on the threshold of a new era, the legacy of past achievements serves as both a foundation and a challenge: to dream bigger, reach further, and embrace the unknown.

The next chapters of this book will delve into how SpaceX and its Starship program are building on this legacy, exploring the technologies, philosophies, and aspirations that are driving humanity's journey into the cosmos. The stars are closer than ever before, and the story of how we get there is just beginning.

Chapter 2

The Basics of Rocket Science

Rockets are more than just marvels of engineering—they are the physical embodiment of humanity's ambition to defy gravity and reach the stars. For millennia, gravity held humanity firmly on the ground, turning the night sky into a canvas of unattainable dreams. The development of rockets transformed that dynamic, enabling us to move beyond the realm of imagination and into the reality of space exploration. However, at the heart of every fiery launch and every orbit lies the intricate dance of physics, engineering, and relentless human determination.

This chapter explores the foundational principles that make rocket science possible, breaking down complex concepts into understandable ideas. From the mechanics of achieving orbit to the intricacies of propulsion systems and the challenges of re-entry, each component plays a vital role in the journey to the stars. To understand the bold vision of programs like SpaceX's Starship, we must first delve into these fundamental building blocks.

The Challenge of Gravity

The first hurdle any rocket must overcome is Earth's gravity, the invisible force that pulls objects toward the planet's center. Gravity keeps everything—from skyscrapers to oceans—anchored to Earth, and it requires tremendous energy to escape its grasp. This gravitational pull is quantified by a value known as escape velocity: the speed an object must reach to break free from Earth's gravitational influence. For Earth, this speed is a staggering 11.2 kilometers per second (about 25,000 miles per hour).

To put this into perspective, consider the energy required to lift even a small object a few feet off the ground. Now scale that up to a spacecraft weighing thousands of tons. The challenge is not just lifting this mass but propelling it fast enough to escape Earth's pull while maintaining stability and direction. This immense energy requirement is why rockets are essentially massive fuel tanks with engines attached—they need an enormous amount of energy to reach space.

Understanding Orbital Mechanics

Achieving orbit isn't just about going straight up. In fact, if a rocket were to travel vertically and then cut its engines, it would fall back to Earth due to gravity. To stay in space, a spacecraft must travel fast enough horizontally to balance the pull of gravity with its forward motion—a delicate equilibrium known as orbital mechanics.

The basic principle is that an object in orbit is continuously falling toward Earth, but because of its horizontal speed, it keeps missing the planet. This concept, first described by Isaac Newton, forms the foundation of orbital mechanics. The required horizontal speed for low Earth orbit (LEO) is about 7.8 kilometers per second (17,500 miles per hour). At this speed, the curvature of the spacecraft's trajectory matches the curvature of Earth, creating a stable orbit.

Orbital mechanics also dictate other key aspects of spaceflight, including the transfer of spacecraft between orbits (such as moving from LEO to a higher orbit or to the Moon), rendezvous and docking with other spacecraft, and re-entry trajectories. Each maneuver requires precise calculations and adjustments, as even minor errors can have catastrophic consequences.

Propulsion Systems: The Engines of Exploration

At the heart of every rocket is its propulsion system, the engine that generates the thrust needed to overcome gravity and reach space. Rocket engines operate on a principle first described by Sir Isaac Newton's third law of motion: for every action, there is an equal and opposite reaction. When a rocket expels exhaust gases downward at high speed, it generates an equal and opposite force that propels the rocket upward.

Rockets use chemical propulsion, where fuel and an oxidizer are burned in a controlled explosion to produce thrust. This reaction occurs in the rocket's combustion chamber, and the high-pressure gases are expelled through a nozzle to create thrust. The type of fuel and oxidizer used can vary, but the two main categories are liquid propulsion and solid propulsion.

Liquid Propulsion Systems:
Liquid rocket engines, like the Raptor engines used by SpaceX, rely on a combination of liquid fuel (such as liquid methane or liquid hydrogen) and a liquid oxidizer (such as liquid oxygen). These components are stored in separate tanks and mixed in the combustion chamber. Liquid engines offer high efficiency and can be throttled, restarted, and controlled with precision, making them ideal for complex missions like landing and reusability.

Solid Propulsion Systems:
Solid rocket engines use a pre-mixed solid fuel and oxidizer packed into a casing. Once ignited, the fuel burns until it is completely consumed, providing a

simple and reliable source of thrust. Solid rockets are often used for initial stages of launch or as boosters, as seen in NASA's Space Shuttle.

Stages of a Rocket's Journey

Rocket launches are divided into stages, each with a specific role in propelling the spacecraft closer to its destination. Multi-stage rockets are used to optimize performance, as carrying spent fuel tanks or engines would waste energy. Each stage is jettisoned when its fuel is depleted, allowing the next stage to continue the journey.

First Stage:
The first stage provides the initial thrust needed to lift the rocket off the ground and through the densest part of Earth's atmosphere. This stage typically contains the largest engines and the majority of the rocket's fuel. In SpaceX's Falcon 9 and Starship systems, the first stage is reusable, landing back on Earth for refurbishment.

Second Stage:
Once the first stage is jettisoned, the second stage takes over, carrying the spacecraft into orbit. This stage is smaller and more efficient, designed for high-altitude performance. It delivers the spacecraft to its target orbit or sets it on a trajectory toward another celestial body.

Payload Stage:
The payload stage carries the mission's objective, whether it's a satellite, scientific instruments, cargo, or astronauts. For interplanetary missions, the payload stage may include additional propulsion systems for deep-space travel.

The Physics of Re-Entry

Re-entry is one of the most critical and dangerous phases of a spacecraft's journey. As a spacecraft descends through Earth's atmosphere, it encounters tremendous friction, generating heat that can reach temperatures of up to 3,000 degrees Fahrenheit (1,650 degrees Celsius). Without proper protection, this intense heat would destroy the spacecraft.

To survive re-entry, spacecraft are equipped with heat shields designed to absorb and dissipate this energy. There are two main types of heat shields: ablative and reusable. Ablative heat shields burn away during re-entry, carrying heat with them, while reusable heat shields, like those on SpaceX's Starship, are designed to withstand multiple flights.

Re-entry dynamics also require precise control to ensure the spacecraft follows the correct trajectory. Too steep an angle could cause the spacecraft to burn up, while too shallow an angle could result in skipping off the atmosphere like a stone on water. The combination of aerodynamic forces, heat management, and navigation makes re-entry a complex and high-stakes operation.

The Importance of Reusability

Traditionally, rockets were single-use machines. Once a mission was completed, the spent stages were discarded, falling into the ocean or burning up in the atmosphere. This approach made space exploration prohibitively expensive, as each mission required an entirely new rocket. Reusability has revolutionized this paradigm, making spaceflight more economical and sustainable.

SpaceX pioneered reusable rocket technology with the development of the Falcon 9, which can land its first stage vertically and be relaunched multiple times. This innovation drastically reduces the cost of launches, enabling more frequent missions and democratizing access to space. The Starship system takes this concept further, with a fully reusable design for both the booster and the spacecraft.

Reusability is not just about cost savings—it's also about creating a sustainable space economy. With reusable systems, it becomes feasible to launch satellites, transport cargo, and conduct deep-space missions at a fraction of the traditional cost. This shift is critical for ambitious goals like colonizing Mars, where frequent and affordable launches will be essential.

The Road Ahead

Rocket science is a field that constantly evolves, driven by the pursuit of greater efficiency, safety, and capability. Each new innovation builds on centuries of scientific discovery, from Newton's laws of motion to the cutting-edge propulsion systems of today. As we look to the future, the principles of rocket science will continue to shape humanity's journey into the cosmos, enabling us to explore new worlds and expand our horizons.

Understanding the basics of rocket science is not just about appreciating the mechanics of spaceflight—it's about recognizing the ingenuity and determination that drive us to reach for the stars. With programs like SpaceX's Starship leading the way, the dream of interplanetary travel is no longer a distant possibility but an impending reality. The rockets of tomorrow will not only carry us further into space but also bring us closer to understanding our place in the universe.

Chapter 3

The SpaceX Revolution

When Elon Musk founded SpaceX in 2002, the aerospace industry seemed as impenetrable as the atmosphere rockets were designed to breach. Dominated by government agencies and a handful of legacy contractors, space exploration was defined by towering budgets, deliberate caution, and incremental progress. Missions required decades of planning, with risk aversion often stifling bold innovation. Musk, however, envisioned something radically different: a future where space travel was fast, affordable, and routine—a future where humanity could extend its reach beyond Earth and establish a permanent presence on other planets.

The founding of SpaceX was not just an entrepreneurial gamble; it was a calculated challenge to an industry weighed down by tradition. Musk's goal was not to play by the established rules but to rewrite them entirely. The company's journey, from its earliest failures to its industry-defining successes, reshaped the world's perception of what was possible. SpaceX introduced the concept of iterative innovation into an industry known for glacial progress, showing that disruption, when paired with bold vision, could lead to transformational change.

This chapter explores the origins of SpaceX, its key milestones, and the revolutionary technologies that have made it a leader in space exploration. We'll examine how SpaceX's approach differs from traditional aerospace companies and explore its impact on global competition in spaceflight.

The Birth of SpaceX

SpaceX was born out of a singular dream: to make life multi-planetary. Musk's vision for humanity's future—thriving on Mars, free from the existential risks of remaining on a single planet—was as ambitious as it was unconventional. At the time, the idea of private companies building rockets and competing with national space agencies seemed outlandish. Yet Musk believed that innovation and cost reduction, fueled by private enterprise, were the keys to opening space for broader participation.

In its early days, SpaceX faced skepticism from both the aerospace industry and the public. Musk personally invested $100 million into the venture, assembling a small team of engineers who shared his belief in the company's mission. Their goal was to create rockets that could be developed quickly, launched affordably, and used repeatedly—a stark departure from the expensive, single-use designs that dominated the industry.

The first major challenge was to design and launch a rocket from scratch. The result was the *Falcon 1*, a two-stage rocket powered by a single Merlin engine. The team worked tirelessly to develop the vehicle, battling limited resources, tight deadlines, and technical hurdles. But the road to success was anything but smooth. Falcon 1's first three launches, conducted between 2006 and 2008, ended in failure, nearly bankrupting the company. Each failure brought hard lessons, but SpaceX persisted, driven by the belief that success was within reach.

The Falcon 1: A Turning Point

The fourth launch of Falcon 1, in September 2008, marked a turning point for SpaceX. This time, the rocket successfully reached orbit, becoming the first privately developed liquid-fueled vehicle to do so. The achievement was more than a technical milestone—it was proof that a private company could compete with national space agencies and legacy contractors. The success of Falcon 1 validated SpaceX's approach, earning the company its first significant contract with NASA under the Commercial Orbital Transportation Services (COTS) program.

Falcon 1 demonstrated that spaceflight didn't have to be the exclusive domain of governments and billion-dollar budgets. It laid the foundation for what would become SpaceX's hallmark: the ability to innovate quickly, learn from failures, and achieve high performance at a fraction of the traditional cost.

The Falcon 9: Reusability Redefined

Building on the lessons of Falcon 1, SpaceX set its sights on a more powerful and versatile rocket: the *Falcon 9*. First launched in 2010, Falcon 9 was designed with two main goals: reliability and reusability. Its two-stage design and modular architecture allowed it to carry a wide range of payloads, from satellites to crewed missions. But what truly set Falcon 9 apart was its potential for reusability—a concept that would revolutionize the aerospace industry.

Traditional rockets were expendable, discarded after a single use. This approach made spaceflight extraordinarily expensive, with the rocket itself accounting for a significant portion of mission costs. SpaceX challenged this paradigm by developing technology to recover and reuse the Falcon 9's first stage.

The idea was simple in concept but incredibly complex in execution: after separating from the second stage, the first stage would return to Earth, using its engines to perform a controlled landing.

The first successful landing of a Falcon 9 booster occurred in December 2015, on a landing pad at Cape Canaveral. It was a historic moment, demonstrating that reusable rockets were not only possible but viable. Over the next several years, SpaceX refined the technology, achieving landings on both solid ground and autonomous drone ships at sea. By 2020, the company had launched and landed the same Falcon 9 booster a record ten times, showcasing the cost savings and operational efficiency of reusability.

The Dragon Capsule: Opening New Horizons

While the Falcon 9 was making headlines, SpaceX was also developing another game-changing vehicle: the *Dragon capsule*. Designed to transport cargo—and later, humans—to the International Space Station (ISS), Dragon became an essential component of SpaceX's partnership with NASA. In 2012, Dragon made history as the first commercial spacecraft to deliver cargo to the ISS, cementing SpaceX's role as a reliable partner in orbital logistics.

The development of the *Crew Dragon* capsule, capable of carrying astronauts, marked another leap forward. In 2020, SpaceX launched its first crewed mission, *Demo-2*, carrying NASA astronauts Doug Hurley and Bob Behnken to the ISS. The mission was a resounding success, making SpaceX the first private company to send humans into orbit. Crew Dragon's sleek design, advanced automation, and reusability set a new standard for crewed spacecraft, further solidifying SpaceX's leadership in human spaceflight.

Starship: The Vision for a Multi-Planetary Future

While Falcon 9 and Dragon revolutionized Earth orbit, Musk's ultimate goal required something far more ambitious: a fully reusable spacecraft capable of carrying humans and cargo to Mars. Enter *Starship*, a next-generation vehicle designed to fulfill SpaceX's vision of interplanetary colonization.

Starship represents a bold departure from traditional rocket design. Unlike Falcon 9, which uses a two-stage configuration, Starship is a fully integrated system with both a reusable booster (Super Heavy) and a reusable spacecraft. Constructed primarily from stainless steel, Starship combines durability, heat resistance, and cost efficiency. Its versatility allows it to support a wide range of missions, from lunar landings to Mars colonization and even Earth-to-Earth point-to-point travel.

The development of Starship has been marked by SpaceX's characteristic approach: rapid iteration and constant testing. Early prototypes, known as Starhopper and SN-series vehicles, underwent rigorous testing, including high-altitude flights and landing attempts. Each test provided valuable data, paving the way for more advanced iterations. Despite dramatic failures—including explosions during landing attempts—SpaceX continued to refine the design, achieving its first successful high-altitude landing in 2021.

Starship is not just a technical challenge; it's a philosophical statement. Its goal is to make space travel affordable and accessible, enabling humanity to expand beyond Earth. For Musk, Starship is the key to a future where humans can thrive as a multi-planetary species, reducing the risks of planetary extinction and unlocking new opportunities for exploration and innovation.

The SpaceX Model: Disruption Through Innovation

SpaceX's success is rooted in its ability to disrupt traditional aerospace models. Unlike legacy contractors, which operate on fixed budgets and rigid timelines, SpaceX embraces a culture of agility, experimentation, and cost efficiency. The company builds its rockets in-house, vertically integrating production to reduce costs and maintain control over the process. It also prioritizes iterative design, using real-world testing to identify and solve problems quickly.

This approach stands in stark contrast to traditional aerospace companies, which often rely on government contracts and focus on minimizing risk. While legacy contractors prioritize perfection, SpaceX embraces failure as a stepping stone to progress. Each failed test is seen as an opportunity to learn and improve, a philosophy that has enabled the company to innovate at an unprecedented pace.

Global Competition and the New Space Race

SpaceX's success has sparked a new era of global competition in space exploration. Traditional aerospace companies like Boeing and Lockheed Martin have been forced to adapt, developing their own reusable rocket technologies to compete with SpaceX. Internationally, countries like China and Russia are accelerating their space programs, aiming to establish their presence in Earth orbit and beyond.

One of the most significant impacts of SpaceX has been its role in lowering the cost of space access. By making launches more affordable, the company has opened the door to new industries, including satellite constellations, space tourism, and lunar exploration. This democratization of space has created a thriving commercial ecosystem, with companies like Blue Origin, Rocket Lab, and Virgin Galactic joining the race.

Legacy and Future of SpaceX

The story of SpaceX is far from over. With each successful mission, the company moves closer to achieving its ultimate goal: making life multi-planetary. From reusable rockets to interplanetary spacecraft, SpaceX's innovations have redefined what is possible in aerospace, inspiring a new generation of dreamers and engineers.

As the world looks to the stars, SpaceX remains at the forefront of the journey. Its legacy is not just about the rockets it builds but about the spirit of exploration it embodies—a spirit that dares to dream bigger, reach further, and redefine humanity's place in the universe.

PART 2

INSIDE STARSHIP'S DEVELOPMENT

Chapter 4

Building the Unbuildable

The idea of building a fully reusable spacecraft capable of interplanetary travel was long regarded as an unattainable dream. It defied conventional wisdom, challenged established engineering principles, and demanded solutions to problems no one had ever solved. Spacecraft had traditionally been disposable, designed for single-use missions where every component was pushed to its limits and then discarded. The notion of creating a spacecraft that could not only withstand the rigors of launch, flight, and re-entry but also be refurbished and flown again was dismissed by many as impractical and unnecessary. Yet, for SpaceX, this vision was not optional; it was a requirement for fulfilling the company's mission of making life multi-planetary.

To realize this dream, SpaceX faced unprecedented technical challenges. The spacecraft would need to endure the searing heat of atmospheric re-entry, survive the harsh conditions of deep space, and operate reliably over multiple missions. It would have to be versatile enough to support a wide range of tasks—from launching satellites and refueling in orbit to landing on Mars and beyond—while remaining cost-efficient and scalable. The result of this ambition is Starship, a revolutionary vehicle designed to push the boundaries of what is possible in aerospace engineering. This chapter explores the immense technical hurdles SpaceX has faced in building Starship, the innovative solutions they have developed, and how they have balanced cost-efficiency with cutting-edge technology.

The Challenge of Reusability

Reusability is the cornerstone of SpaceX's strategy for reducing the cost of spaceflight. Traditionally, rockets were discarded after a single use, making each mission prohibitively expensive. For Musk's vision of affordable and frequent interplanetary travel to become a reality, this model had to be overturned. The idea of reusability is deceptively simple: build a spacecraft that can fly, land, and be refurbished for another mission. But achieving this requires overcoming a cascade of engineering challenges.

One of the most significant obstacles is the stress that rockets endure during flight. From the moment of liftoff, the vehicle experiences extreme vibrations, aerodynamic forces, and rapid changes in temperature. The most intense challenge comes during re-entry, where the spacecraft encounters atmospheric friction that generates temperatures of up to 3,000 degrees Fahrenheit. Designing a vehicle that can survive this environment and remain functional for subsequent missions demands materials and systems that are not only durable but also lightweight and cost-effective.

SpaceX's first foray into reusability began with the Falcon 9 rocket, whose first stage was designed to return to Earth and land vertically. While this represented a major breakthrough, the Falcon 9's second stage and payload fairings remained expendable. Starship, in contrast, aims to be fully reusable, with both the booster (Super Heavy) and the spacecraft designed for multiple flights. This holistic approach to reusability marks a dramatic departure from traditional aerospace practices.

The Raptor Engine: Powering the Future

At the heart of Starship's design is the Raptor engine, a technological marvel that represents one of the most advanced rocket engines ever built. Raptor engines are powered by liquid methane and liquid oxygen, a combination chosen for its efficiency and compatibility with long-term space missions. Unlike traditional engines that use kerosene-based fuels, the choice of methane aligns with SpaceX's vision for Mars colonization, as methane can potentially be produced on

Mars using the planet's carbon dioxide and water resources.

The Raptor engine operates on a full-flow staged combustion cycle, a design that maximizes efficiency by burning all of the propellant in the engine. This approach generates extremely high pressure, resulting in a thrust-to-weight ratio that surpasses most traditional rocket engines. Each Raptor engine produces 500,000 pounds of thrust, and Starship will use a cluster of up to 33 Raptors on its Super Heavy booster and six on the spacecraft itself.

Developing the Raptor engine posed significant engineering challenges. The high pressures involved in the full-flow staged combustion cycle require components capable of withstanding extreme stress and heat. To address this, SpaceX employed advanced manufacturing techniques, including 3D printing, to produce critical components with precision and efficiency. The iterative testing of Raptor prototypes, conducted at SpaceX's facilities in Texas, allowed engineers to rapidly refine the design and address failures, embodying the company's philosophy of learning through experimentation.

Stainless Steel: A Material Revolution

One of the most striking aspects of Starship's design is its gleaming stainless-steel body. While most modern rockets are built from lightweight composites or aluminum alloys, SpaceX opted for stainless steel—a choice that initially raised eyebrows within the aerospace community. However, the decision was driven by practicality and innovation.

Stainless steel offers several advantages for a reusable spacecraft. First, it has excellent heat resistance, allowing it to endure the intense temperatures of atmospheric re-entry without the need for extensive thermal protection systems. Second, it is durable and easy to work with, making it suitable for rapid manufacturing and repair. Third, stainless steel is cost-effective compared to specialized aerospace materials, aligning with SpaceX's goal of reducing costs.

The specific alloy used in Starship, known as 300-series stainless steel, is optimized for strength, corrosion resistance, and thermal performance. The material's ability to withstand high temperatures eliminates the need for traditional ablative heat shields, which burn away during re-entry. Instead, Starship employs a combination of stainless steel and reusable ceramic tiles in key areas, creating a thermal protection system that can be refurbished for multiple missions.

Heat Shields: Surviving Re-Entry

Re-entry is one of the most perilous phases of any mission, as spacecraft face extreme heat and aerodynamic forces. For Starship, designing a heat shield capable of surviving multiple re-entries was a critical challenge. Traditional heat shields, such as those used on NASA's Apollo capsules, are ablative, meaning they sacrifice material to absorb and dissipate heat. While effective, this approach is inherently single-use.

SpaceX took a different approach, developing a reusable heat shield system for Starship. The spacecraft's underbelly is covered in hexagonal ceramic tiles that protect the vehicle from the intense heat of re-entry. These tiles are designed to withstand temperatures far higher than those experienced during launch, ensuring the spacecraft remains intact as it plunges through Earth's atmosphere at hypersonic speeds.

The development of these heat shields has been a process of trial and error. Early prototypes faced challenges with tile adhesion and durability, but iterative testing allowed engineers to refine the design. The use of stainless steel in the spacecraft's structure also reduces the reliance on heat shields, as the material itself can tolerate high temperatures. This integrated approach ensures both safety and reusability.

Balancing Cost-Efficiency with Innovation

One of SpaceX's defining characteristics is its ability to balance cost-efficiency with cutting-edge technology. Traditional aerospace projects often involve long development timelines and inflated budgets, driven by a focus on minimizing risk. SpaceX, in contrast, embraces a philosophy of rapid iteration, prioritizing speed and learning over perfection.

This approach is evident in the company's manufacturing processes. By vertically integrating production, SpaceX controls every aspect of its supply chain, reducing costs and ensuring quality. Advanced manufacturing techniques, such as 3D printing, enable the rapid production of complex components, while modular designs allow for easy assembly and testing. The use of stainless steel, with its lower material costs and ease of fabrication, further contributes to affordability.

Innovation is also driven by SpaceX's willingness to accept failure as part of the development process. Starship prototypes, such as the SN-series vehicles, have undergone a series of dramatic tests, including high-altitude flights and landing attempts. While some tests ended in fiery explosions, each provided valuable data that informed subsequent iterations. This iterative approach has allowed SpaceX to make rapid progress, achieving milestones in a fraction of the time typical of traditional aerospace projects.

Testing and Iteration: The Path to Progress

The development of Starship has been characterized by a relentless cycle of testing, failure, and improvement. Early prototypes, such as Starhopper and the SN-series vehicles, were used to validate key systems, including propulsion, aerodynamics, and landing techniques. Each test, whether successful or not, offered insights that were incorporated into the next iteration.

One of the most ambitious tests occurred in May 2021, when the Starship SN15 prototype successfully completed a high-altitude flight and landing. This marked a significant milestone in the program, demonstrating the feasibility of Starship's design and the effectiveness of its systems. Subsequent prototypes have continued to refine the vehicle, paving the way for orbital tests and eventual interplanetary missions.

SpaceX's iterative testing process reflects its broader philosophy: to learn by doing. This approach, while risky, has enabled the company to achieve breakthroughs that would have been impossible with a more conservative strategy.

The Vision Behind Starship

Starship is more than just a spacecraft; it is the embodiment of SpaceX's vision for humanity's future. Musk has often described Starship as a "Holy Grail" project, capable of transforming space exploration and making life multi-planetary. Its versatility allows it to support a wide range of missions, from launching satellites and refueling in orbit to establishing colonies on Mars and beyond.

The vehicle's modular design and full reusability make it uniquely suited for long-term space missions. Its stainless-steel body and methane-fueled engines enable it to operate in diverse environments, from the vacuum of space to the thin atmosphere of Mars. With a capacity to carry over 100 tons of cargo, Starship has the potential to revolutionize industries such as space tourism, planetary science, and deep-space exploration.

For Musk, Starship represents a solution to the existential risks facing humanity. By establishing a presence on other planets, we can ensure the survival of our species in the face of threats such as climate change, asteroid impacts, and geopolitical conflict. Starship is not just a technical achievement; it is a symbol of hope and resilience.

The Unbuildable, Built

Building Starship has required SpaceX to confront some of the most complex challenges in aerospace engineering. From the development of the Raptor engine to the design of reusable heat shields and the use of stainless steel, each innovation represents a leap forward in technology and capability. Through its bold vision and relentless pursuit of progress, SpaceX has turned the dream of a fully reusable interplanetary spacecraft into a tangible reality.

Starship is more than a spacecraft; it is a testament to the power of human ingenuity and ambition. It embodies the belief that no challenge is insurmountable, no dream too big. As SpaceX continues to refine and expand the capabilities of Starship, it is paving the way for a future where humanity's reach extends far beyond the confines of Earth. The unbuildable has been built, and the journey to the stars has only just begun.

Chapter 5

Trial by Fire

Space exploration has always been a domain where success is hard-won and failure is an integral part of the process. For SpaceX, the development of Starship has been no exception. The path to building a fully reusable spacecraft capable of interplanetary travel has been paved with fiery tests, hard lessons, and relentless perseverance. Each test flight, each explosion, and each milestone has contributed valuable data, helping SpaceX refine its designs and move closer to its ultimate goal of making life multi-planetary.

This chapter delves into the dramatic journey of Starship's testing program, from its earliest prototypes to its most significant breakthroughs. It examines the critical moments that defined the program, the challenges SpaceX faced, and the lessons learned along the way. Through trial by fire, the Starship program has become a testament to SpaceX's philosophy of rapid iteration and the power of resilience in the face of adversity.

Early Prototypes: From Starhopper to SN Series

The journey began with *Starhopper*, an unassuming prototype that resembled a water tower more than a spacecraft. Starhopper was not designed to reach orbit; instead, its purpose was to test the feasibility of the Raptor engine and validate basic flight and landing systems. Its first tethered flight in April 2019 was a modest hop, lifting the vehicle a few meters off the ground. While it was far from the grandeur of a full orbital mission, it marked an important step forward.

In August 2019, Starhopper completed its first untethered flight, soaring 150 meters into the air before making a controlled landing. This simple demonstration provided critical insights into the Raptor engine's performance and the precision of SpaceX's landing technology. Though its flights were short and low, Starhopper proved the foundational concepts behind Starship were sound, paving the way for more advanced prototypes.

The next phase of development introduced the *SN series* (Serial Number), each iteration pushing the boundaries of what Starship could achieve. The SN prototypes were designed for high-altitude flight tests, with the goal of testing aerodynamic control, engine performance, and landing maneuvers. These vehicles were taller, sleeker, and more capable than Starhopper, featuring the distinctive stainless-steel body that would become Starship's hallmark.

High-Altitude Flight Tests: Learning Through Failure

The first major test of the SN series came with *SN8* in December 2020. The prototype was equipped with three Raptor engines and was tasked with a high-altitude flight to 12.5 kilometers (approximately 41,000 feet). The launch was a dramatic success; SN8 ascended smoothly, shutting down its engines sequentially to simulate orbital flight conditions. It then executed a "belly flop" maneuver, flipping onto its side to test Starship's aerodynamic surfaces during descent. This maneuver was crucial for ensuring the spacecraft could maintain stability during re-entry and slow its descent without relying solely on propulsion.

However, the landing attempt ended in a fiery explosion. SN8 failed to reduce its velocity sufficiently, resulting in a hard impact. While some might view this as a failure, SpaceX considered it a major success. The test provided invaluable data on engine performance, aerodynamic control, and landing dynamics, all of which were used to refine subsequent prototypes.

The pattern continued with *SN9* and *SN10*, both of which conducted similar high-altitude tests. SN9, launched in February 2021, successfully completed its flight and belly flop but also suffered a crash landing. SN10, launched just weeks later, achieved the program's first successful landing—only to explode minutes later due to a fuel tank issue. Each test was a spectacle, capturing global attention and solidifying SpaceX's reputation for resilience and innovation.

By the time *SN15* launched in May 2021, the program had reached a new level of maturity. This prototype incorporated numerous upgrades based on the lessons learned from earlier tests. SN15 successfully completed its high-altitude flight and landed safely, marking a significant milestone for the Starship program. It demonstrated that the vehicle's design was robust and capable of achieving its goals, providing the foundation for the next phase of testing.

The Evolution of the Heat Shield

One of the most critical components of Starship's design is its *heat shield*, which protects the spacecraft during atmospheric re-entry. Re-entry is an extraordinarily challenging phase of any mission, as the spacecraft encounters intense friction with the atmosphere, generating temperatures of up to 3,000 degrees

Fahrenheit. For Starship, this challenge is magnified by its goal of reusability; the heat shield must not only survive re-entry but also remain functional for multiple flights.

SpaceX's solution is a system of *hexagonal ceramic tiles* that cover the spacecraft's underbelly. These tiles are designed to absorb and dissipate heat, shielding the stainless-steel structure beneath. However, developing this system has been a complex and iterative process.

Early prototypes faced challenges with tile adhesion and durability. During some tests, tiles were seen detaching from the spacecraft, exposing the underlying structure to potential damage. Engineers worked tirelessly to address these issues, experimenting with new materials, adhesives, and attachment methods. They also conducted rigorous ground tests, subjecting the tiles to simulated re-entry conditions to evaluate their performance.

By the time of SN15, the heat shield had undergone significant improvements, demonstrating its ability to withstand the stresses of flight and landing. Future iterations of Starship will continue to refine this system, ensuring it can handle the even greater demands of interplanetary missions.

Key Breakthroughs and Milestones

The first major breakthrough in Starship's testing program came with *SN15*, which proved that the vehicle could complete a high-altitude flight and landing without catastrophic failure. This success validated the design and marked a turning point for the program, shifting the focus toward more ambitious goals.

Another critical milestone was the first test of the *Super Heavy booster*, the massive first stage of the Starship system. Equipped with up to 33 Raptor engines, Super Heavy is designed to provide the thrust needed to lift Starship into orbit. Initial static fire tests demonstrated the booster's immense power, while subsequent flight tests will evaluate its performance during launch and landing.

The program also achieved a significant milestone with the first *orbital flight attempt*, which aimed to demonstrate Starship's ability to reach space and return safely. While this test highlighted areas for improvement, it underscored SpaceX's commitment to learning through iteration and pushing the boundaries of what is possible.

The Philosophy of Iteration

At the heart of SpaceX's testing program is a philosophy of *rapid iteration*. Unlike traditional aerospace projects, which prioritize perfection and risk avoidance, SpaceX embraces failure as a necessary step toward progress. Each test, whether successful or not, provides valuable data that informs the next iteration. This approach allows SpaceX to move quickly, making continuous improvements and achieving milestones at an unprecedented pace.

The company's willingness to take risks and learn from failure has become a defining characteristic of its culture. It has enabled SpaceX to innovate in ways that were previously thought impossible, from landing reusable rockets to developing the world's most powerful engines. The Starship program exemplifies this philosophy, demonstrating that even the most ambitious goals can be achieved through perseverance and adaptation.

Lessons Learned and Future Goals

The trials and triumphs of the Starship testing program have yielded countless lessons. Engineers have gained insights into propulsion, aerodynamics, heat management, and structural integrity, all of which are critical for the success of future missions. These lessons have not only advanced the Starship program but have also contributed to the broader field of aerospace engineering.

Looking ahead, SpaceX's goals for Starship are nothing short of revolutionary. The vehicle is expected to play a central role in NASA's Artemis program, which aims to return humans to the Moon. It will also serve as a platform for Mars colonization, enabling the transportation of people and cargo to the Red Planet. Beyond these goals, Starship has the potential to revolutionize industries such as space tourism, satellite deployment, and deep-space exploration.

Triumph Through Trial

The development of Starship has been a journey defined by bold ambition, relentless testing, and an unwavering commitment to progress. Through each trial, whether marked by success or failure, SpaceX has moved closer to realizing its vision of a fully reusable spacecraft capable of interplanetary travel. The program's major tests have not only demonstrated the feasibility of its design but have also inspired a new generation of dreamers and innovators.

As Starship continues to evolve, it serves as a reminder that the path to greatness is rarely smooth. It is through trial by fire that humanity advances, forging new paths and reaching for the stars. The lessons of the Starship program will shape the future of space exploration, proving that even the most daunting challenges can be overcome with determination, creativity, and the courage to try again.

Chapter 6

Iteration and Innovation

At the heart of SpaceX's extraordinary progress lies a philosophy that has redefined aerospace engineering: rapid iteration and the embrace of failure as a catalyst for learning. In an industry where caution and perfection have traditionally reigned, SpaceX's approach of "failing fast and learning faster" has not only accelerated innovation but also shattered long-held assumptions about what is possible in space exploration. This philosophy, embedded deeply in the company's culture, has enabled SpaceX to achieve in years what once took decades.

Iteration and innovation are more than just buzzwords for SpaceX—they are a mindset, a process, and a mission. Every prototype, every test flight, and every explosion serves as a stepping stone toward progress, transforming setbacks into opportunities for discovery. This chapter delves into the philosophy of rapid prototyping, the role of failure in accelerating progress, and how unexpected outcomes have shaped Starship's design and mission.

The Philosophy of Rapid Prototyping

SpaceX's approach to engineering is built on the principle of rapid prototyping. Rather than perfecting a design before testing it, the company adopts a cycle of designing, building, testing, and iterating at an unprecedented pace. This process allows engineers to identify and resolve issues quickly, incorporating lessons from each iteration into the next version. The result is a continuous improvement cycle that drives progress faster than traditional methods.

In traditional aerospace programs, designs are often refined and reviewed for years before they are ever built or tested. The goal is to minimize risk and avoid failure at all costs, but this cautious approach can lead to stagnation and missed opportunities for innovation. SpaceX, by contrast, accepts failure as an inevitable and necessary part of development. Each failed test is viewed as a data-rich experiment, offering insights that are impossible to glean from simulations or theoretical models alone.

This philosophy was evident from the earliest days of the Starship program. Instead of waiting to build a perfect prototype, SpaceX began with simple test vehicles like Starhopper, which was little more than a steel tank with a single engine. These early prototypes were crude but functional, providing the company with critical data on engine performance, landing stability, and aerodynamic control. By starting small and iterating rapidly, SpaceX was able to refine its designs in real time, achieving progress at a pace that few in the industry thought possible.

The Role of Failure in Accelerating Progress

Failure, far from being a setback, is a cornerstone of SpaceX's success. The company's culture encourages engineers to take risks, push boundaries, and view failures as opportunities for learning. This mindset has enabled SpaceX to tackle complex challenges head-on, often achieving breakthroughs that traditional approaches would have missed.

One of the most dramatic examples of this philosophy in action is the series of high-altitude flight tests conducted with Starship prototypes. Beginning with SN8 in December 2020, these tests aimed to validate key systems, including engine performance, aerodynamic stability, and landing mechanisms. While each test was ambitious, none of the early flights ended perfectly. SN8 and SN9 both experienced successful ascents and belly-flop maneuvers but crashed during their landing attempts. SN10 managed to land but exploded shortly after touchdown due to a fuel tank issue.

To an outside observer, these failures might have seemed like disasters. But for SpaceX, they were invaluable experiments. Each test revealed weaknesses in the design, providing engineers with precise data on what went wrong and how to fix it. By embracing failure as a tool for learning, SpaceX was able to iterate rapidly, achieving a successful high-altitude flight and landing with SN15 just months later.

This cycle of testing, failing, and improving is not unique to Starship. It has been a defining feature of every major SpaceX project, from the Falcon 1 rocket to the Falcon 9's reusable booster. Each failure brings the company closer to its goal, reinforcing the idea that innovation is not about avoiding mistakes but about learning from them.

Unexpected Outcomes Shaping Starship's Design

In the process of building Starship, SpaceX has encountered numerous unexpected outcomes that have fundamentally shaped the vehicle's design. These moments of serendipity, born out of failure or unanticipated challenges, have often led to breakthroughs that redefined the program's direction.

1. The Belly-Flop Maneuver:

One of the most iconic aspects of Starship's design is its belly-flop maneuver, which involves flipping the spacecraft onto its side during descent to slow its fall using aerodynamic drag. While this approach was planned, the execution revealed unforeseen complexities. During early tests, engineers discovered that maintaining stability during the maneuver required more precise control of the vehicle's flaps and engines than initially anticipated. This led to the development of more sophisticated software algorithms and real-time adjustments to improve Starship's stability and maneuverability.

The data from these tests also influenced the placement and design of Starship's flaps, optimizing them for both atmospheric entry and landing. The belly-flop maneuver, while challenging to perfect, has become a defining feature of Starship, enabling it to re-enter Earth's atmosphere and land without relying on excessive propulsion.

2. Heat Shield Development:

The evolution of Starship's heat shield is another example of how unexpected outcomes have driven innovation. Early prototypes faced issues with the adhesive used to attach ceramic tiles, leading to tiles detaching during tests. These failures prompted SpaceX to experiment with new attachment methods, ultimately developing a more reliable system that ensured tiles would remain intact during flight.

The process also revealed insights into the behavior of heat shield materials under extreme conditions. Engineers discovered that certain areas of the spacecraft experienced higher thermal loads than expected, leading to localized tile reinforcements. These lessons have been critical for ensuring the heat shield's durability during atmospheric re-entry, a key requirement for Starship's reusability.

3. Raptor Engine Optimization:
The development of the Raptor engine, which powers both the Starship and its Super Heavy booster, has also benefited from iterative testing. During early engine tests, engineers encountered issues with combustion stability and high-pressure performance. By analyzing the failures, they were able to refine the engine's design, improving its efficiency and reliability.

One notable outcome was the decision to use a full-flow staged combustion cycle, a complex but highly efficient method of burning propellant. While challenging to implement, this approach has allowed SpaceX to achieve higher thrust-to-weight ratios and better performance under extreme conditions. The iterative improvements to the Raptor engine have made it one of the most advanced rocket engines in the world, capable of supporting the ambitious goals of the Starship program.

Scaling Innovation: From Prototype to Production

As Starship moves from prototype testing to operational missions, SpaceX faces the challenge of scaling its innovation processes for production. Rapid iteration is effective in the development phase, but building a fleet of operational Starships requires a balance between experimentation and standardization.

SpaceX has addressed this challenge by adopting a modular approach to manufacturing. Each component of Starship, from its engines to its heat shield tiles, is designed for mass production, enabling the company to scale up quickly. Advanced manufacturing techniques, such as 3D printing, play a critical role in this process, allowing engineers to produce complex parts with precision and efficiency.

The company's production facility in Boca Chica, Texas, has become a hub of innovation, with multiple prototypes under construction at any given time. This parallel production approach allows SpaceX to test new designs while building operational vehicles, ensuring a steady pipeline of progress. The facility itself is designed for scalability, with plans to produce dozens of Starships annually to support a wide range of missions.

Cultural Innovation: A Mindset for the Future

The success of SpaceX's iterative approach is not just about technology—it's about culture. The company's engineers are encouraged to take risks, think creatively, and challenge assumptions. This mindset fosters a spirit of innovation that permeates every aspect of the organization, from design and testing to manufacturing and operations.

Musk himself has been a driving force behind this culture, often pushing teams to aim higher and move faster than they thought possible. While this approach can be demanding, it has created an environment where breakthroughs are celebrated and failures are seen as opportunities for growth. This culture of resilience and ambition is what sets SpaceX apart, enabling it to tackle challenges that others might deem insurmountable.

The Broader Impact of Iteration and Innovation

SpaceX's philosophy of iteration and innovation has had a profound impact on the aerospace industry as a whole. By demonstrating that rapid prototyping and risk-taking can lead to faster progress, the company has inspired other organizations to rethink their approaches to development. Traditional aerospace companies, once focused on perfection and predictability, are beginning to adopt more agile methodologies, embracing the idea that failure is a necessary part of progress.

This shift is not limited to the aerospace sector. SpaceX's success has influenced industries ranging from automotive manufacturing to software development, proving that iterative design can drive innovation across a wide range of fields. The lessons learned from Starship's development are shaping the future of engineering, inspiring a new generation of innovators to dream bigger and move faster.

The Power of Iteration

The story of Starship is a testament to the power of iteration and innovation. By embracing failure, learning from mistakes, and continuously refining its designs,

SpaceX has achieved breakthroughs that once seemed impossible. The company's rapid prototyping philosophy has enabled it to overcome challenges, push boundaries, and redefine what is possible in space exploration.

As Starship moves closer to operational readiness, it serves as a reminder that progress is not a straight line. It is a journey of trial and error, of bold experiments and unexpected discoveries. Through iteration and innovation, SpaceX is not only building a spacecraft—it is building a future where humanity can thrive among the stars.

Chapter 7

Hot Staging and Re-Entry

Re-entry is one of the most perilous and technologically demanding phases of any spacecraft's journey. A returning vehicle must endure extreme heat, intense aerodynamic forces, and precisely controlled trajectories to ensure a safe landing. For SpaceX's Starship, the challenge of re-entry is magnified by its ambitious goals: not only does the spacecraft need to survive this fiery ordeal, but it must do so in a way that preserves its reusability for multiple flights. This requirement transforms re-entry from a one-off survival task into a repeatable process, pushing the boundaries of aerospace engineering.

Starship's re-entry strategy represents a radical departure from traditional approaches. It leverages advanced materials, cutting-edge thermal protection systems, and a dramatic "belly-flop" maneuver to slow its descent. Each test flight has provided valuable insights into the science and mechanics of re-entry, enabling SpaceX to refine its systems and prepare for the challenges of future missions to Mars. This chapter explores the intricate dynamics of Starship's re-entry process, the innovations that make it possible, and how these developments are shaping the future of interplanetary travel.

The Science of Re-Entry: Facing the Heat

When a spacecraft returns to Earth, it must decelerate from orbital speeds of approximately 28,000 kilometers per hour (17,500 miles per hour). This dramatic reduction in velocity is achieved through the friction generated as the vehicle collides with atmospheric particles. While this friction is essential for slowing the spacecraft, it also creates immense heat, with temperatures around the vehicle reaching as high as 3,000 degrees Fahrenheit (1,650 degrees Celsius). Without robust thermal protection, the spacecraft would be destroyed during this phase.

The heat generated during re-entry occurs due to *compression heating*, where the spacecraft compresses the air in front of it, creating a shockwave. The energy from this shockwave is transferred to the surrounding air, turning it into a superheated plasma. This plasma, while visually stunning, poses significant risks to the spacecraft's structure and onboard systems.

Traditional spacecraft like NASA's Apollo capsules relied on *ablative heat shields*, which burned away during re-entry, carrying the heat with them. While effective, this approach rendered the heat shield single-use. Starship, however, is designed to be fully reusable, necessitating a more advanced thermal protection system.

Starship's Thermal Protection System

Starship's re-entry strategy centers on its innovative *thermal protection system (TPS)*. The spacecraft's underbelly is covered with hexagonal ceramic tiles, which are designed to withstand extreme temperatures and protect the stainless-steel structure beneath. These tiles, while deceptively simple in appearance, are the product of extensive research and development.

Each ceramic tile is crafted to absorb and dissipate heat, preventing it from reaching the spacecraft's internal components. The hexagonal shape minimizes gaps between tiles, reducing the risk of hot spots where heat could penetrate. Additionally, the tiles are attached using a robust adhesive system that has been refined through iterative testing. Early prototypes faced issues with tiles detaching during flight, but SpaceX's engineers have addressed these challenges through improved materials and attachment methods.

The choice of *stainless steel* for Starship's structure also plays a crucial role in its thermal management. Unlike traditional aluminum-based alloys, stainless steel can withstand higher temperatures, providing an additional layer of protection during re-entry. This synergy between the heat shield tiles and the underlying structure ensures that Starship can survive the intense conditions of re-entry and be refurbished for future flights.

The Aerodynamics of Re-Entry: The Belly-Flop Maneuver

One of the most striking aspects of Starship's design is its *belly-flop maneuver*, a dramatic and unconventional approach to re-entry. Unlike traditional spacecraft, which descend nose-first or base-first, Starship flips onto its side during re-entry, presenting its broad underbelly to the atmosphere. This orientation maximizes aerodynamic drag, slowing the spacecraft's descent without relying solely on propulsion.

The belly-flop maneuver is a key component of Starship's re-entry strategy for several reasons. First, it reduces the reliance on engines, conserving fuel for other mission phases. Second, the wide surface area of the spacecraft's underbelly allows it to dissipate heat more effectively, minimizing the thermal load on any single point. Finally, the maneuver enables precise control of the spacecraft's trajectory and orientation, allowing it to target specific landing zones.

Executing the belly-flop maneuver requires precise coordination between Starship's aerodynamic surfaces and its Raptor engines. The spacecraft is equipped with four large flaps—two at the front and two at the rear—that can be independently adjusted to maintain stability during descent. These flaps act as control surfaces, enabling Starship to make subtle adjustments to its angle of attack and descent rate.

The Innovation of Hot Staging

In addition to its re-entry capabilities, Starship incorporates another innovative feature: *hot staging*, a technique that allows the vehicle to transition seamlessly between flight phases. Hot staging involves igniting the engines of the second stage while it is still attached to the first stage, creating a smoother and more efficient separation process.

Hot staging offers several advantages for Starship's design. By igniting the second stage engines before separation, the vehicle maintains continuous thrust, minimizing the loss of velocity during the transition. This efficiency is particularly important for achieving orbital trajectories, where every meter per second of velocity is critical. Additionally, hot staging reduces the mechanical complexity of the separation process, enhancing reliability and safety.

SpaceX has conducted extensive tests to validate the feasibility of hot staging for Starship. These tests have demonstrated the system's ability to maintain stability and control during the transition, providing a foundation for future orbital missions. The integration of hot staging into Starship's design reflects SpaceX's commitment to optimizing every aspect of its spacecraft, from launch to landing.

Starlink-Enabled Telemetry: Real-Time Insights

One of the most groundbreaking innovations in Starship's testing program is the use of *Starlink-enabled telemetry*, which provides real-time data on the spacecraft's performance during re-entry. Traditional telemetry systems often experience communication blackouts during re-entry due to the plasma generated by the shockwave. Starlink, however, bypasses this limitation by using a network of low-Earth orbit satellites to maintain continuous connectivity.

This real-time telemetry has been invaluable for SpaceX's engineers, allowing them to monitor critical systems during flight and identify areas for improvement. By analyzing data on heat distribution, aerodynamic forces, and structural integrity, engineers can make precise adjustments to Starship's design. The integration of Starlink also underscores SpaceX's ability to leverage its ecosystem of technologies to enhance its testing and development processes.

Preparing for Mars: The Ultimate Test

The lessons learned from Starship's re-entry tests are not just about returning to Earth—they are about preparing for the challenges of interplanetary travel. Mars presents a unique set of obstacles for re-entry, landing, and surface operations, requiring systems that can adapt to its thin atmosphere and extreme temperatures.

Mars' atmosphere is approximately 1% as dense as Earth's, meaning that aerodynamic braking is far less effective. Starship's belly-flop maneuver, which relies on maximizing drag, will need to be adapted for this environment. Additionally, the spacecraft's heat shield will need to withstand the unique thermal conditions of Martian re-entry, where lower atmospheric density results in different heating profiles.

SpaceX has designed Starship with these challenges in mind, conducting tests that simulate the conditions of Mars landings. The development of autonomous landing systems, capable of identifying safe landing zones and adjusting trajectories in real time, will be critical for ensuring mission success. These systems, combined with the lessons learned from Earth-based re-entry tests, will pave the way for Starship's future role as a workhorse for Mars colonization.

The Future of Re-Entry Technology

The innovations pioneered by SpaceX in the Starship program have the potential to revolutionize re-entry technology for the entire aerospace industry. By demonstrating the feasibility of reusable thermal protection systems, advanced aerodynamics, and integrated telemetry, SpaceX is setting new standards for spacecraft design. These developments could be applied to a wide range of missions, from commercial satellite launches to deep-space exploration.

As Starship continues to evolve, its re-entry systems will undergo further refinement, incorporating lessons from each test flight. The ultimate goal is to create a spacecraft that can operate reliably and sustainably across multiple missions, reducing the cost and complexity of space exploration. In doing so, SpaceX is not only advancing its own capabilities but also driving innovation across the industry, inspiring a new generation of engineers and explorers.

Pioneering the Path to Mars

The re-entry phase of spaceflight has always been a formidable challenge, demanding the perfect balance of science, engineering, and precision. For Starship, this challenge is magnified by its dual goals of reusability and interplanetary travel. Through its innovative thermal protection systems, aerodynamic maneuvers, and advanced telemetry, SpaceX is pushing the boundaries of what is possible, laying the groundwork for a future where spacecraft can travel between planets as easily as they orbit Earth.

As the Starship program moves closer to operational readiness, its re-entry systems will play a crucial role in enabling the next era of space exploration. Whether returning to Earth or landing on Mars, Starship's ability to navigate the fiery crucible of re-entry will determine the success of its missions—and, ultimately, humanity's ability to thrive beyond our home planet. Through hot staging, real-time telemetry, and the relentless pursuit of innovation, SpaceX is charting a path toward the stars, proving that even the most daunting challenges can be overcome with vision, determination, and courage.

PART 3

THE HUMAN ELEMENT

Chapter 8

The People Behind Starship

While the gleaming metal of Starship and the roaring engines of its boosters capture the world's imagination, the real power behind SpaceX lies in its people. The engineers, scientists, and visionaries who labor tirelessly to bring the impossible to life are the unsung heroes of the Starship program. Their work is fueled not only by technical expertise but also by a culture of collaboration, resilience, and relentless innovation. At the center of this culture is Elon Musk, whose audacious vision and leadership have shaped the company's trajectory.

This chapter explores the lives and contributions of the key individuals behind Starship, providing a glimpse into the unique culture that drives SpaceX. It also examines how Musk's vision has influenced the program, inspiring a team of thousands to tackle some of the most complex engineering challenges in history. Starship is not just a testament to technology—it's a testament to human ingenuity, perseverance, and the power of shared ambition.

Elon Musk: The Architect of the Dream

Elon Musk's vision for humanity is as bold as it is unconventional. At the heart of his philosophy is the belief that humanity must become a multi-planetary species to ensure its long-term survival. From the beginning, this goal has been the driving force behind SpaceX, and Starship is its ultimate manifestation. Musk's willingness to think beyond conventional boundaries has not only shaped the company's direction but also redefined the global aerospace industry.

Musk's hands-on leadership style sets him apart from traditional CEOs. He is deeply involved in the technical aspects of SpaceX, regularly engaging with engineers and scientists to solve complex problems. His ability to grasp the intricacies of rocket science, combined with his relentless drive, has helped SpaceX achieve milestones that once seemed impossible. Musk is known for setting audacious goals, such as landing humans on Mars within the next decade, and pushing his team to achieve them with unwavering focus.

However, Musk's leadership style is not without controversy. His demanding expectations and relentless pace have been criticized by some as overly intense. Yet, for many at SpaceX, his vision serves as a source of inspiration. Employees often describe Musk as someone who challenges them to reach their full potential, pushing the boundaries of what they thought was achievable. Under his leadership, SpaceX has become a hub for some of the brightest minds in engineering and science, all united by a shared commitment to exploration and innovation.

Gwynne Shotwell: The Steady Hand

While Elon Musk is the public face of SpaceX, much of the company's success can be attributed to its president and COO, Gwynne Shotwell. A mechanical engineer by training, Shotwell joined SpaceX in 2002 as the company's seventh employee. Over the years, she has played a pivotal role in transforming SpaceX from a fledgling startup into a global leader in aerospace.

Shotwell is widely respected for her ability to balance the technical and business aspects of SpaceX's operations. She has been instrumental in securing contracts with NASA and private customers, ensuring the company's financial stability and growth. Her calm and pragmatic approach complements Musk's visionary leadership, creating a dynamic partnership that has propelled SpaceX to new heights.

Under Shotwell's guidance, SpaceX has developed a reputation for reliability and innovation. She has been a key advocate for Starship, helping to secure its role in

NASA's Artemis program and positioning it as a cornerstone of future space exploration. Shotwell's leadership exemplifies the collaborative and mission-driven culture that defines SpaceX, inspiring employees to work toward a shared goal of making space accessible to all.

The Engineers and Scientists Behind Starship

The Starship program is a testament to the ingenuity and dedication of thousands of engineers and scientists, each contributing their expertise to overcome seemingly insurmountable challenges. These individuals work across a wide range of disciplines, from propulsion and materials science to software development and aerodynamics. While their names may not be widely known, their contributions are integral to the program's success.

Propulsion Experts: The Power Behind the Raptor Engine
The development of the Raptor engine, one of the most advanced rocket engines ever built, is a feat of engineering brilliance. Led by Tom Mueller, SpaceX's former VP of Propulsion, and later continued by a team of propulsion experts, the Raptor engine team tackled the immense challenges of creating a full-flow staged

combustion engine powered by liquid methane and oxygen. This design maximizes efficiency and performance, enabling Starship to achieve its ambitious goals.

Each Raptor engine is a masterpiece of precision, capable of withstanding extreme pressures and temperatures. The propulsion team's work extends beyond the engines themselves, encompassing the fuel systems, igniters, and nozzles that make Starship's launches possible. Their iterative approach to testing and refinement has been critical in ensuring the reliability and safety of the engines.

Thermal Protection Innovators: The Guardians Against Heat

Starship's ability to survive re-entry depends on its thermal protection system, a complex network of ceramic tiles that shields the spacecraft from temperatures exceeding 3,000 degrees Fahrenheit. The team responsible for this system includes materials scientists, thermal engineers, and manufacturing experts who have worked tirelessly to perfect the design.

Early challenges with tile detachment and durability required innovative solutions, including advanced adhesives and novel attachment methods. The team's rigorous testing, often conducted under extreme conditions, has ensured that the heat shield can withstand multiple re-entries without compromising the spacecraft's integrity.

Aerodynamic Specialists: Masters of the Belly-Flop

The iconic belly-flop maneuver, which allows Starship to slow its descent using aerodynamic drag, is the result of meticulous work by SpaceX's aerodynamic specialists. These engineers analyze airflow, control surface dynamics, and structural stress to optimize the spacecraft's performance during re-entry. Their work has been critical in ensuring that Starship can maintain stability and control during its descent, even in the most challenging conditions.

The Culture at SpaceX: Collaboration and Resilience

SpaceX's culture is as unique as its technology. The company thrives on collaboration, bringing together experts from diverse disciplines to solve complex problems. Teams work in close proximity, sharing ideas and feedback in real time. This flat organizational structure fosters innovation, allowing employees at all levels to contribute to the development process.

Resilience is another defining characteristic of SpaceX's culture. The company's willingness to embrace failure as a learning opportunity has created an environment where employees feel empowered to take risks and push boundaries. Each test flight, whether successful or not, is viewed as a step toward progress, reinforcing the idea that setbacks are an inevitable part of achieving ambitious goals.

SpaceX's culture also emphasizes speed and efficiency. Engineers are encouraged to iterate rapidly, focusing on solving problems rather than getting bogged down in bureaucracy. This approach has enabled SpaceX to achieve milestones at an unprecedented pace, from landing reusable boosters to launching the first private spacecraft to dock with the International Space Station.

The Impact of Elon Musk's Vision

Elon Musk's vision has had a profound impact on the Starship program, shaping its goals, culture, and technological trajectory. His belief in humanity's potential to become a multi-planetary species has inspired a sense of purpose that extends beyond the confines of SpaceX. For many employees, working on Starship is not just a job—it's a mission to advance the future of exploration.

Musk's focus on cost reduction and reusability has also driven SpaceX's engineering philosophy. By prioritizing affordability and scalability, the company has made space exploration more accessible, paving the way for a new era of innovation and collaboration. This commitment to democratizing space is evident in every aspect of Starship's design, from its modular construction to its reusable heat shields.

Musk's influence extends beyond SpaceX, inspiring a global shift in how the aerospace industry approaches innovation. His willingness to challenge conventional thinking has sparked a wave of competition and collaboration, encouraging other companies and agencies to pursue bold ideas and take greater risks.

The Future of Starship's Team

As Starship moves closer to operational readiness, the team behind the program continues to grow and evolve. New challenges, such as developing in-orbit refueling and preparing for Mars missions, require fresh perspectives and expertise. SpaceX's ability to attract top talent from around the world ensures that it remains at the forefront of innovation.

The lessons learned from the Starship program are also influencing future projects, from lunar landers to deep-space exploration vehicles. The engineers and scientists who have contributed to Starship's success are shaping the next generation of spacecraft, applying their knowledge and experience to push the boundaries of what is possible.

The Human Factor in Space Exploration

The story of Starship is not just a story of technology—it is a story of people. The engineers, scientists, and leaders behind the program are the driving force behind its success, turning dreams into reality through their ingenuity, dedication, and collaboration. From Elon Musk's visionary leadership to the tireless efforts of the propulsion, thermal protection, and aerodynamic teams, each individual has played a critical role in bringing Starship to life.

As humanity looks to the stars, the people behind Starship remind us that exploration is a collective endeavor, one that requires the combined talents of countless individuals working toward a shared goal. Their work is not only shaping the future of space exploration but also inspiring a new generation to reach for the stars. Through their efforts, Starship has become more than a spacecraft—it has become a symbol of what humanity can achieve when we dare to dream big and work together.

Chapter 9

The Public's Spacecraft

Space exploration has always captured the imagination of humanity, transcending borders, generations, and cultural divides. The Apollo Moon landings inspired millions around the globe, while the Voyager missions gave us awe-inspiring glimpses of the outer planets. SpaceX's Starship program is the latest chapter in this legacy of wonder, but with a modern twist. Unlike the space programs of the past, Starship has become a uniquely public endeavor—one that invites the world to witness its progress, share in its triumphs, and participate in the dream of exploring the cosmos.

Through live broadcasts, social media, and unprecedented transparency, SpaceX has built a global community of enthusiasts, dreamers, and future explorers. Starship is not just a spacecraft—it is a cultural phenomenon, a source of inspiration for millions, and a vehicle for rekindling humanity's curiosity about the universe. This chapter delves into how Starship has captured the public's imagination, the role of media in building excitement around space exploration, and how this engagement is inspiring a new generation to reach for the stars.

The Power of Inspiration: A Global Dream

The idea of humanity traveling to other planets has long been a staple of science fiction, but SpaceX's Starship program has turned it into a tangible reality. By openly sharing its vision of interplanetary colonization, SpaceX has made the dream of life on Mars something people can believe in, discuss, and aspire to. This openness has helped Starship transcend the realm of engineering and become a symbol of possibility.

Starship has captured the public's imagination in part because of its audacity. The idea of a fully reusable spacecraft that can carry 100 tons of cargo—or dozens of humans—to another planet is inherently thrilling. The spacecraft's sleek, futuristic design evokes images of classic science fiction, from the rocket ships of *Flash Gordon* to the starships of *Star Wars*. But unlike fiction, Starship is real, and its progress is visible to the world.

This sense of accessibility is key to Starship's appeal. People from all walks of life can watch its development unfold in real time, marveling at test flights, analyzing successes and failures, and dreaming about what the future might hold. For many, Starship represents more than a technological achievement; it is a reminder of humanity's potential to overcome challenges and explore the unknown.

Live Broadcasts: Sharing the Journey

One of the defining features of the Starship program is its commitment to sharing every step of the journey with the public. SpaceX's live broadcasts of test flights have become must-watch events, drawing millions of viewers from around the globe. These broadcasts combine technical insight with a sense of drama and spectacle, creating a shared experience that brings people together.

The broadcasts are notable for their transparency. Viewers see the launches in real time, complete with commentary from SpaceX engineers who explain the objectives, challenges, and technical details of each mission. Unlike traditional aerospace companies, which often keep testing behind closed doors, SpaceX embraces the risks of public failure. When a test ends in a fiery explosion, as many Starship prototypes have, it is not hidden or downplayed. Instead, it is presented as a natural part of the iterative design process, helping the audience understand the challenges of building a spacecraft.

This openness has fostered a sense of trust and engagement between SpaceX and its audience. By allowing people to witness both the successes and setbacks of the Starship program, SpaceX has demystified the process of space exploration, making it more relatable and inspiring. The live broadcasts have turned test flights into global events, sparking conversations about science, engineering, and the future of humanity.

Social Media: Building a Community

Social media has played a pivotal role in making Starship a cultural phenomenon. Platforms like Twitter, YouTube, and Instagram have allowed SpaceX to connect directly with millions of fans, providing updates, answering questions, and sharing behind-the-scenes glimpses of the program. Elon Musk's own Twitter account has become a hub for Starship enthusiasts, where he shares progress updates, technical details, and even memes that reflect the excitement and challenges of the program.

The accessibility of social media has democratized the conversation around space exploration. Fans from all over the world can share their thoughts, create fan art, and engage with SpaceX engineers and other enthusiasts. This sense of community has transformed Starship from a corporate project into a shared dream, one that belongs to everyone who imagines a future among the stars.

Social media has also amplified the voices of young people, many of whom see Starship as an inspiration for pursuing careers in science, technology, engineering, and mathematics (STEM). Through hashtags, viral videos, and online discussions, the excitement around Starship has spread to classrooms, sparking curiosity and ambition among the next generation of scientists and engineers.

Transparency and Engagement: Redefining Public Relations

SpaceX's approach to public engagement represents a fundamental shift in how aerospace companies interact with the public. Historically, space exploration was often shrouded in secrecy, with government agencies and contractors providing limited information about their work. SpaceX, by contrast, has embraced transparency as a core value, sharing its successes and failures with the world.

This transparency has humanized the Starship program, making it clear that progress is not a straight line but a series of iterative steps. When a prototype explodes during a test flight, SpaceX explains what happened, what went wrong, and what lessons were learned. This openness helps the public appreciate the complexity of space exploration and builds trust in the company's commitment to safety and innovation.

SpaceX's engagement strategy extends beyond its live broadcasts and social media presence. The company has hosted public tours of its facilities, invited students to watch launches, and collaborated with educators to create resources that explain the science behind its missions. These efforts have made Starship more than a spacecraft—it has become a tool for education and inspiration, igniting a passion for exploration in people of all ages.

Inspiring a New Generation

The impact of Starship on young people cannot be overstated. For many, watching a Starship test flight is a formative experience, one that sparks curiosity about science, engineering, and the universe. This inspiration is especially important in an era where STEM education is critical for addressing global challenges and advancing human knowledge.

Educators have embraced Starship as a teaching tool, using its test flights and technical details to illustrate concepts in physics, mathematics, and engineering.

From explaining the physics of re-entry to discussing the challenges of building reusable rockets, Starship provides real-world examples that make science come alive for students. The program has also inspired a wave of student-led projects, competitions, and initiatives aimed at exploring the possibilities of space exploration.

The cultural impact of Starship extends beyond classrooms. It has inspired artists, writers, and filmmakers to imagine new stories about humanity's future in space. Fan art depicting Starship on Mars, poems about interplanetary travel, and documentaries about the program have become part of the broader narrative, reflecting how deeply the spacecraft has resonated with the public.

The Role of Failure: A Shared Journey

One of the most compelling aspects of the Starship program is its willingness to embrace failure as part of the process. SpaceX's live broadcasts and social media updates often highlight the risks and uncertainties of test flights, showing the public that setbacks are not just inevitable but necessary for progress. This openness about failure has changed how people perceive space exploration, shifting the narrative from one of perfection to one of perseverance.

When a Starship prototype explodes during a test, it becomes a shared moment for SpaceX and its audience. Fans analyze the footage, discuss the technical details, and celebrate the lessons learned. This collective engagement transforms failure into a communal experience, reinforcing the idea that exploration is a journey of discovery rather than a quest for immediate success.

By sharing its failures and triumphs, SpaceX has created a culture of resilience that extends beyond its workforce to the broader public. The Starship program has become a symbol of the power of persistence, inspiring people to tackle challenges in their own lives with the same determination and optimism.

A Vision for the Future

Starship's ability to capture the public's imagination is not just about what it represents today—it's about what it promises for the future. The idea of humans living and working on Mars, of building cities on other planets, and of exploring the far reaches of the solar system is profoundly inspiring. Starship has turned these dreams into tangible goals, sparking conversations about humanity's place in the universe and the steps we need to take to get there.

For many, Starship is a symbol of hope, a reminder that humanity is capable of achieving great things when we work together and think boldly. It represents the best of what we can be: innovative, curious, and unafraid to explore the unknown. By engaging the public in its journey, SpaceX has made Starship more than a spacecraft—it has made it a beacon of possibility for people around the world.

A Spacecraft for All

Starship is not just a technological marvel; it is a cultural phenomenon that has reignited humanity's passion for space exploration. Through live broadcasts, social media, and transparency, SpaceX has invited the world to be part of its journey, turning Starship into a shared dream. The program has inspired a new generation of scientists, engineers, and explorers, showing them that the stars are not as far away as they once seemed.

As Starship moves closer to operational readiness, its impact on the public will only grow. It is a reminder that space exploration is not just about rockets and technology—it is about people, imagination, and the enduring human desire to discover what lies beyond. Through its openness, resilience, and vision, Starship has become the public's spacecraft, a symbol of what we can achieve when we dare to dream big and reach for the stars.

Chapter 10

Ethical and Philosophical Questions

The prospect of becoming a multi-planetary species, driven by advancements like SpaceX's Starship, is one of humanity's most profound ambitions. It challenges not just our technological capabilities but also our ethical, legal, and philosophical frameworks. As private companies lead the charge in space exploration, questions arise about ownership, governance, and the environmental costs of venturing beyond Earth. At the same time, the idea of expanding humanity's reach raises deeper reflections on our responsibilities to our home planet, our future, and the universe.

This chapter delves into the critical questions surrounding space exploration, examining the ethical dilemmas, environmental impacts, and philosophical considerations of becoming a spacefaring civilization. From debates over who owns space to the existential implications of life on other planets, these issues are as important as the technological breakthroughs making space exploration possible.

Who Owns Space? The Legal and Ethical Implications

Space has long been considered a global commons, a domain that belongs to all humanity rather than any one nation or entity. This principle is enshrined in the *Outer Space Treaty of 1967*, which prohibits the appropriation of celestial bodies by sovereign states and mandates the use of space for peaceful purposes. However, the treaty was drafted during the Cold War, when space exploration was dominated by governments, and it provides limited guidance on the role of private companies in space activities.

With the rise of companies like SpaceX, Blue Origin, and others, the legal landscape of space is becoming increasingly complex. These companies are not just launching satellites—they are planning missions to the Moon, Mars, and beyond, raising questions about resource extraction, territorial claims, and governance. For example, if SpaceX establishes a Mars colony, who has jurisdiction over it? Can a private company claim resources mined from asteroids or the Martian surface? And what legal framework ensures that these activities benefit humanity as a whole?

Private Enterprise and Public Good
One of the key ethical questions is how to balance the interests of private enterprise with the public good. Companies like SpaceX operate within a capitalist framework, driven by profit motives and market dynamics. While their innovations have democratized access to space and accelerated progress, there is a risk that unchecked commercialization could lead to inequities, exploitation, and conflict.

Some argue that space exploration should remain a public endeavor, guided by international cooperation and oversight. Others believe that private companies, with their efficiency and entrepreneurial spirit, are better suited to drive progress. A potential solution lies in a hybrid model, where governments and private entities collaborate under a shared framework that ensures transparency, accountability, and equitable access to space resources.

Space Governance: The Need for New Treaties
The existing treaties governing space, including the Outer Space Treaty and the Moon Agreement, are increasingly inadequate for addressing the complexities of modern space activities. There is a growing need for new international agreements that account for private companies, resource extraction, and long-term habitation of other planets.

Proposals for new governance structures include the establishment of an international regulatory body, akin to the United Nations, to oversee space activities and resolve disputes. This body could set guidelines for resource sharing, environmental protection, and human rights in space. However, achieving consensus among nations—and ensuring compliance from private entities—will be a significant challenge.

The Environmental Impact of Space Travel

While space exploration holds immense promise, it also comes with significant environmental costs. Rockets require vast amounts of fuel, and their launches release greenhouse gases, particulate matter, and other pollutants into the atmosphere. The increasing frequency of launches, driven by commercial demand and scientific missions, raises concerns about the long-term sustainability of these activities.

Rocket Emissions and Climate Change

Rocket launches produce emissions that can have localized and global environmental effects. For example, the combustion of rocket fuel releases carbon dioxide, water vapor, and other gases into the upper atmosphere. These emissions can contribute to the depletion of the ozone layer and exacerbate climate change.

SpaceX has taken steps to address these concerns through the development of the *Raptor engine*, which uses methane and liquid oxygen as propellants. This choice reduces the production of soot and other harmful byproducts compared to kerosene-based fuels. Additionally, methane can be synthesized from renewable resources, offering a pathway to carbon-neutral rocket operations. However, even with these advancements, the environmental footprint of frequent launches remains a challenge.

Space Debris: A Growing Threat
Another environmental concern is the accumulation of *space debris*, or "space junk," in Earth's orbit. Dead satellites, spent rocket stages, and fragments from collisions pose risks to operational spacecraft and the International Space Station. With the proliferation of satellite constellations, such as SpaceX's Starlink, the risk of debris-related incidents is increasing.

SpaceX has implemented measures to mitigate its impact, including designing satellites that deorbit at the end of their operational lives. The company also collaborates with international organizations to track and manage space debris. However, addressing the broader issue will require global cooperation, stricter regulations, and innovative solutions such as debris removal technologies.

Balancing Exploration and Conservation

The environmental costs of space exploration raise a broader ethical question: how do we balance our drive to explore the universe with our responsibility to protect Earth? For SpaceX, this balance is central to its mission. By developing sustainable technologies and prioritizing reusability—such as the ability to relaunch the same Starship multiple times—the company aims to minimize its environmental footprint while advancing its vision of interplanetary travel.

Philosophical Reflections: Becoming a Multi-Planetary Species

The idea of becoming a multi-planetary species is as much a philosophical question as it is a technological one. It challenges us to consider our place in the universe, our responsibilities to each other, and the implications of expanding beyond Earth.

Why Leave Earth?

One of the most compelling arguments for interplanetary travel is the need to safeguard humanity's future. Elon Musk has often cited the existential risks facing our species, including climate change, pandemics, and asteroid impacts. Establishing a presence on other planets, he argues, would act as a "backup" for civilization, ensuring that humanity could survive even in the face of global catastrophes.

However, this rationale raises questions about our priorities. Should we focus on fixing the problems on Earth before venturing into space? Critics argue that the resources spent on space exploration could be better used to address poverty, inequality, and environmental degradation. Proponents counter that the technologies developed for space exploration often have significant benefits for life on Earth, from advances in renewable energy to innovations in medicine and materials science.

The Ethics of Colonization
The term "colonization" carries historical baggage, evoking memories of exploitation, displacement, and inequality. As we consider the possibility of colonizing Mars or other celestial bodies, it is essential to approach this endeavor with a commitment to equity and inclusion. Who decides who gets to go to Mars? How do we ensure that these new societies are governed fairly and sustainably? These questions are critical for avoiding the mistakes of the past and building a future that reflects our highest ideals.

Our Place in the Cosmos
Becoming a multi-planetary species also invites us to reflect on our place in the cosmos. For centuries, humanity has gazed at the stars and wondered about our

role in the universe. The idea of living on other planets challenges us to think beyond the boundaries of Earth, imagining new possibilities for culture, identity, and meaning.

Philosophers have long debated the implications of space exploration for our understanding of humanity. Some view it as an extension of our innate curiosity and drive to explore, while others see it as a moral imperative to preserve life and consciousness in the universe. Regardless of perspective, the pursuit of interplanetary travel forces us to confront profound questions about who we are, what we value, and what kind of future we want to create.

Space Exploration as a Reflection of Humanity

At its core, space exploration is a reflection of humanity's greatest strengths and weaknesses. It embodies our curiosity, ingenuity, and resilience, as well as our capacity for ambition and cooperation. But it also highlights the challenges we face, from managing our impact on the environment to addressing issues of equity and governance.

The Starship program, with its bold vision and groundbreaking technology, represents both the promise and the complexity of this endeavor. It inspires us to think big, take risks, and imagine a future where humanity thrives on multiple planets. But it also challenges us to consider the ethical, environmental, and philosophical dimensions of this journey, reminding us that exploration is not just about where we go but how we go and why.

The Questions That Define Us

As humanity stands on the cusp of a new era in space exploration, the ethical and philosophical questions raised by programs like Starship are as important as the technological breakthroughs that make them possible. Who owns space? How do we protect the environment while pursuing our ambitions? What does it mean to become a multi-planetary species? These questions challenge us to think deeply about our responsibilities, our values, and our vision for the future.

SpaceX's Starship program has brought these questions to the forefront, sparking conversations that will shape the next chapter of human history. By addressing these challenges with thoughtfulness and care, we can ensure that our journey to the stars reflects the best of who we are—and who we aspire to be. Through exploration, we not only expand our horizons but also deepen our understanding of what it means to be human.

PART 4

BROADER IMPLICATIONS

Chapter 11

Economics of the Final Frontier

Space exploration, once the exclusive domain of governments with vast budgets, has become an arena of entrepreneurial innovation, driven by private companies like SpaceX. At the heart of this transformation lies a fundamental question: How do you make space travel economically sustainable? The answer, as demonstrated by SpaceX, involves rethinking traditional financial models, embracing reusability, and leveraging cost-efficient technologies to unlock new opportunities for exploration and commerce.

As humanity sets its sights on interplanetary colonization and commercial ventures beyond Earth, the economics of the final frontier become increasingly complex. From the costs of launching rockets to the challenges of establishing self-sustaining colonies on Mars, each step of the journey requires a careful balance of ambition and practicality. This chapter examines the financial model of SpaceX, explores the economic challenges and opportunities of interplanetary colonization, and considers how commercial space ventures could shape the future of humanity.

The Financial Model of SpaceX: Revolutionizing Cost Efficiency

SpaceX's success is built on a radical rethinking of how rockets are designed, built, and operated. Traditional aerospace companies operate on a cost-plus model, where contracts with governments guarantee profits regardless of cost overruns. SpaceX, by contrast, operates more like a lean startup, focusing on efficiency, innovation, and market-driven strategies.

Reusability as the Cornerstone
One of SpaceX's most significant innovations is the development of reusable rockets, which drastically reduce the cost of space travel. Traditionally, rockets were single-use machines, discarded after launching their payloads into orbit. This approach made space travel prohibitively expensive, with launch costs reaching tens or even hundreds of millions of dollars per mission.

SpaceX's Falcon 9 rocket changed the game. By designing the first stage to return to Earth and land vertically, SpaceX introduced a new paradigm: rockets that could be flown multiple times with minimal refurbishment. This innovation cut launch costs by a factor of ten, making space more accessible for commercial and scientific missions. The Starship program takes this concept further, aiming for full reusability of both the booster and spacecraft, which could reduce costs even more dramatically.

Vertical Integration
Another key to SpaceX's cost-efficiency is its vertically integrated supply chain. Unlike traditional aerospace companies, which outsource components to multiple suppliers, SpaceX designs and manufactures most of its rocket parts in-house. This approach not only reduces costs but also gives the company greater control over quality and innovation. For example, the Raptor engines used in Starship are built entirely by SpaceX, allowing the company to iterate quickly and optimize performance.

Economies of Scale
SpaceX's ambitious launch cadence also plays a critical role in its financial model. By launching rockets more frequently than any other company or government agency, SpaceX achieves economies of scale, spreading fixed costs over a larger number of missions. This high-volume approach allows the company to offer competitive prices for satellite launches, making it a preferred choice for commercial customers.

Diversified Revenue Streams
While SpaceX's launch services are a major source of revenue, the company has diversified its business to include other ventures, such as the Starlink satellite internet network. Starlink, which aims to provide global high-speed internet coverage, represents a potentially massive market, with billions of dollars in annual revenue. This financial cushion enables SpaceX to fund its more ambitious projects, such as the development of Starship and Mars colonization efforts.

The Economics of Interplanetary Colonization

Colonizing another planet is one of the most audacious goals in human history, and it comes with astronomical costs. From launching spacecraft to building habitats, every aspect of interplanetary colonization presents economic challenges that must be addressed to make the endeavor feasible.

Transportation Costs
The first major hurdle is the cost of transporting people and cargo to another planet. While SpaceX's Starship is designed to make interplanetary travel more affordable, each mission will still require significant resources, including fuel, food, and life-support systems. Reducing

these costs will be critical for enabling large-scale colonization.

SpaceX aims to address this challenge through innovations such as *in-orbit refueling*, which allows spacecraft to carry less fuel at launch and refuel in space for the journey to Mars. Additionally, the use of *methane-based propulsion systems* is key to the long-term sustainability of Mars missions, as methane can be produced on Mars using local resources.

Self-Sustaining Economies
A successful colony on Mars must eventually become self-sustaining to reduce its reliance on Earth. This requires the development of local industries, such as agriculture, manufacturing, and energy production. Establishing these industries will involve significant upfront investment in infrastructure, including power plants, greenhouses, and mining equipment.

One potential economic model for Mars colonization is the creation of a "frontier economy," where early settlers contribute to building the colony while generating revenue through mining, research, and tourism. Over time, as the population grows and industries develop, the colony could achieve greater self-sufficiency, reducing the need for financial support from Earth.

The Role of Private Investment

Funding large-scale colonization efforts will require significant private investment, in addition to government support. Companies and individuals with an interest in the economic opportunities of space could play a critical role in financing infrastructure, technology development, and early missions. SpaceX's success in attracting investors for ventures like Starlink demonstrates the potential for private capital to drive progress in space exploration.

Commercial Space Ventures: Opportunities and Challenges

As space becomes more accessible, a wide range of commercial ventures are emerging, from satellite deployment to space tourism. These ventures represent both opportunities and challenges for the future of space exploration.

Satellite Markets and Space Infrastructure
Satellite deployment is currently one of the most lucrative segments of the space industry. Companies and governments rely on satellites for telecommunications, navigation, weather monitoring, and Earth observation. SpaceX has capitalized on this market with its reliable and cost-effective launch services, as well as its own Starlink constellation.

Beyond satellites, the development of space infrastructure—such as space stations, refueling depots, and manufacturing facilities—could unlock new economic opportunities. These projects will require collaboration between governments, private companies, and international organizations to ensure their feasibility and sustainability.

Space Tourism
Space tourism, while still in its infancy, has the potential to become a major industry. Companies like Blue Origin and Virgin Galactic are already offering suborbital flights for wealthy customers, and SpaceX has announced plans for private missions to the Moon and beyond. While the high costs of space tourism currently limit its accessibility, advancements in technology and economies of scale could eventually make it more affordable for a broader audience.

Asteroid Mining and Resource Extraction
Asteroid mining is another frontier of commercial space activity, with the potential to generate trillions of dollars in economic value. Asteroids contain vast quantities of valuable resources, including rare metals, water, and other materials that could support both space exploration and terrestrial industries. While the technology for asteroid mining is still in development, companies like Planetary Resources and Deep Space Industries are exploring its potential.

However, the pursuit of these resources raises ethical and legal questions, including concerns about environmental impact, ownership, and equitable access. Developing a regulatory framework for space resource extraction will be essential for ensuring that these activities benefit humanity as a whole.

The Long-Term Vision: A Spacefaring Economy

As humanity moves closer to becoming a spacefaring civilization, the economics of space will continue to evolve. The transition from Earth-based economies to interplanetary economies will require new financial models, innovative technologies, and collaborative governance.

The Role of Governments and International Collaboration

Governments will play a critical role in supporting the development of space economies, providing funding for scientific research, infrastructure, and regulatory oversight. International collaboration will be essential for addressing shared challenges, such as space debris, resource management, and planetary protection.

Expanding Markets and Industries

The growth of space economies could lead to the emergence of entirely new markets and industries, from space-based energy production to interplanetary trade. These developments will create opportunities for innovation, investment, and economic growth, while also raising questions about inequality, labor rights, and the distribution of wealth.

The Ethical Dimension
The economic expansion into space also carries ethical implications. How do we ensure that the benefits of space exploration are shared equitably? How do we balance profit motives with the need to protect the environment and preserve the cultural and scientific value of celestial bodies? These questions will shape the policies and practices of space economies in the decades to come.

The Economics of Exploration

The economics of space exploration are as challenging and dynamic as the technological advancements that drive it. From SpaceX's innovative financial model to the complexities of interplanetary colonization, every aspect of the journey to the final frontier requires a careful balance of vision, strategy, and practicality.

SpaceX's success in making space travel more affordable and accessible has opened the door to a new era of exploration and commerce. As humanity ventures further into the cosmos, the development of sustainable economic systems will be essential for ensuring that our activities in space benefit not just a few but all of humanity. By addressing these challenges with creativity and collaboration, we can build an economic foundation that supports our dreams of exploring the stars and thriving as a multi-planetary species.

Chapter 12

SpaceX and the New Space Race

The first Space Race, ignited by Cold War rivalries between the United States and the Soviet Union, captivated the world with its triumphs and tragedies. It gave us milestones like the launch of Sputnik, the first human in orbit, and the Apollo Moon landings, which forever transformed humanity's relationship with space. Decades later, the dynamic of space exploration has shifted dramatically. What was once a duel between superpowers is now a complex, multi-faceted arena that includes a mix of national agencies, private companies, and international collaborations.

At the forefront of this new era is SpaceX, a private company that has not only challenged the dominance of traditional government space programs but also set new standards for innovation, cost-efficiency, and ambition. Its achievements have spurred a modern "New Space Race," pitting nations and companies against one another in a race for dominance in the final frontier. This chapter explores how SpaceX's accomplishments compare with international efforts, examines the geopolitical implications of a privatized space race, and considers what this new competition means for the future of humanity.

The Rise of SpaceX: A Catalyst for Change

SpaceX has redefined what is possible in space exploration. From reusable rockets to plans for Mars colonization, the company's achievements have forced governments and competitors to rethink their strategies. Central to SpaceX's success is its ability to achieve goals once thought impossible—quickly, cost-effectively, and often with greater ambition than traditional agencies.

Key Achievements of SpaceX
1. Reusable Rockets: SpaceX's Falcon 9 revolutionized spaceflight by introducing reusable first-stage boosters, dramatically reducing the cost of launches. The ability to land and reuse these boosters upended decades of aerospace tradition and set a new benchmark for efficiency.

2. Crewed Missions: In 2020, SpaceX's Crew Dragon became the first privately built spacecraft to carry astronauts to the International Space Station (ISS), marking a historic milestone in commercial space exploration.

3. Starlink: SpaceX has deployed thousands of satellites for its Starlink program, which aims to provide global high-speed internet coverage. This constellation is not only a commercial success but also a demonstration of SpaceX's capability to scale operations rapidly.

4. Starship Development: SpaceX's Starship program, with its vision of fully reusable spacecraft for interplanetary travel, represents the boldest leap in aerospace technology. Its potential applications range from lunar landings to Mars colonization, making it a cornerstone of SpaceX's ambition.

These milestones have placed SpaceX at the forefront of the New Space Race, challenging not only traditional aerospace companies but also national space agencies.

International Efforts in the New Space Race

While SpaceX has captured the public's imagination, it operates in a highly competitive global landscape. National space agencies like NASA, the European Space Agency (ESA), and China's CNSA, as well as private competitors like Blue Origin, are advancing their own programs. Each brings unique strengths and strategies to the race.

NASA: A Partner and Competitor

NASA, SpaceX's most prominent partner and occasional competitor, has been reinvigorated by its collaboration with private companies. Programs like the Commercial Crew Program, which funds companies to develop spacecraft for crewed missions, have enabled NASA to focus on ambitious goals like the Artemis program.

The Artemis program aims to return humans to the Moon by 2025 and establish a sustainable presence on its surface. NASA has selected SpaceX's Starship as the lunar lander for the mission, a testament to the company's technological capabilities. However, NASA continues to pursue its own developments, such as the Space Launch System (SLS), a heavy-lift rocket designed for deep-space exploration. Despite delays and cost overruns, SLS represents NASA's commitment to maintaining leadership in space exploration.

European Space Agency (ESA): Collaboration and Independence

The ESA plays a significant role in the New Space Race, emphasizing international collaboration and cutting-edge science. Its programs, such as the ExoMars mission and the Galileo satellite navigation system, highlight its focus on scientific exploration and infrastructure.

While ESA collaborates with NASA and SpaceX on various projects, it also seeks to maintain independence.

Recent initiatives, like the Ariane 6 rocket, aim to compete with SpaceX in the commercial launch market. However, ESA faces challenges in matching SpaceX's cost-efficiency and rapid innovation.

China: A Rising Space Power
China's space program, led by the China National Space Administration (CNSA), has emerged as a formidable force. With ambitious plans for lunar exploration, a space station, and Mars missions, China has positioned itself as a key player in the New Space Race.

China's achievements include the Chang'e lunar program, which successfully landed rovers on the Moon, and the Tianwen-1 mission, which placed a rover on Mars in 2021. The CNSA is also constructing the Tiangong space station, which aims to rival the ISS as a hub for scientific research and international collaboration.

Unlike SpaceX, China's space program is government-led, with significant state resources and long-term strategic planning. However, China's rise has introduced geopolitical tensions, particularly with the United States, as both nations vie for influence in space.

Other Players
Other countries, such as Russia, India, and Japan, also play important roles in the New Space Race. Russia's

Roscosmos has a storied history in space exploration, but its dominance has waned in recent years due to economic challenges and competition from private companies. India's ISRO has made impressive strides, including the Chandrayaan lunar missions and its low-cost Mars Orbiter Mission, demonstrating the potential for cost-effective space exploration. Japan's JAXA continues to lead in scientific missions, such as the Hayabusa asteroid-sampling program, and collaborates with international partners on large-scale projects.

Geopolitical Implications of a Privatized Space Race

The privatization of space exploration has profound geopolitical implications. While the first Space Race was driven by national pride and Cold War rivalries, the New Space Race is a complex interplay of government interests, private enterprise, and international cooperation.

Shifting Power Dynamics
The rise of companies like SpaceX has shifted the balance of power in space exploration. Traditionally, space was the domain of nation-states, with government agencies setting the agenda and controlling access to space. Today, private companies wield significant influence, often outpacing national programs in

innovation and efficiency. This shift raises questions about the role of governments in regulating and supporting space activities.

International Collaboration vs. Competition

While space has historically been a domain of collaboration—exemplified by the ISS and joint scientific missions—the New Space Race introduces new competitive dynamics. Countries and companies compete for leadership in key areas, such as lunar exploration, satellite deployment, and Mars colonization. This competition can drive innovation but also risks creating tensions, particularly in the absence of clear international regulations.

The Militarization of Space

Another concern is the militarization of space. The establishment of the U.S. Space Force and similar initiatives in other countries signal growing recognition of space as a strategic domain. Satellites play a critical role in communication, navigation, and surveillance, making them targets in potential conflicts. The development of anti-satellite weapons and other military technologies raises the stakes in the New Space Race, highlighting the need for agreements to prevent the escalation of conflict in space.

The Role of SpaceX in the New Space Race

SpaceX occupies a unique position in the New Space Race, serving as both a partner to national agencies and a disruptor of traditional models. Its success has spurred other players to innovate, creating a ripple effect across the industry.

Driving Innovation
SpaceX's rapid development cycle and focus on reusability have set new standards for the industry. Competitors are now racing to match its cost-efficiency and technological advancements, from reusable rockets to commercial crewed missions. This competition benefits humanity as a whole, accelerating progress and reducing costs for access to space.

Shaping Policy and Regulation
As a private company, SpaceX operates within a complex regulatory framework that balances innovation with public interest. Its activities, from Starlink deployments to Mars colonization plans, raise questions about resource management, environmental impact, and governance. SpaceX's prominence ensures that its actions will influence the development of policies and treaties governing space exploration.

Inspiring Public Engagement

SpaceX's openness and transparency have also inspired public interest in space exploration. By broadcasting launches, sharing progress on Starship, and engaging with fans on social media, SpaceX has turned space exploration into a global conversation. This renewed enthusiasm has the potential to drive support for space programs worldwide, fostering a sense of shared purpose in the exploration of the cosmos.

A New Era of Exploration

The New Space Race is a dynamic and evolving competition, driven by the ambitions of governments, private companies, and international partnerships. SpaceX's achievements have set a high bar, challenging traditional models and inspiring a new wave of innovation. At the same time, the rise of international players like China and the continued leadership of NASA and ESA highlight the diverse approaches shaping this era of exploration.

As humanity ventures further into the cosmos, the New Space Race raises profound questions about cooperation, competition, and the shared future of space exploration. Will this race lead to greater collaboration and mutual benefit, or will it deepen geopolitical divides? The answer will depend on how governments, companies, and citizens navigate the challenges and opportunities of this new frontier.

In the end, the New Space Race is not just about technological achievement—it is about defining humanity's place in the universe. By embracing both competition and collaboration, we can ensure that this race is not only a contest but also a journey of discovery that benefits all of humanity.

Chapter 13

Preparing for Mars

The idea of colonizing Mars has captivated humanity for centuries, but it is only in the past few decades that it has begun to transition from science fiction to a tangible goal. The Red Planet represents the ultimate challenge: a destination that pushes the boundaries of our technological, logistical, and human capabilities. SpaceX, with its bold mission to make life multi-planetary, has placed Mars at the center of its vision, with Starship as the cornerstone for this audacious endeavor.

Colonizing Mars is not just about engineering a spacecraft capable of reaching another planet—it's about solving an interconnected web of challenges that span propulsion, life support, energy generation, and human adaptability. This chapter explores the myriad obstacles that must be overcome to establish a permanent presence on Mars, the critical role of Starship in this mission, and how SpaceX is preparing for humanity's most ambitious journey yet.

Why Mars? The Case for Colonization

Mars has long been considered the most viable option for human colonization beyond Earth. Its proximity, surface conditions, and resources make it uniquely suited for long-term habitation compared to other celestial bodies in our solar system.

Proximity and Accessibility
Mars is the closest planet to Earth that offers the potential for sustained human presence. While Venus is closer during certain alignments, its extreme surface temperatures and atmospheric pressures make it uninhabitable. Mars, on the other hand, has a day length similar to Earth's, seasonal variations, and surface gravity that is about 38% of Earth's—enough to support human physiology over extended periods.

Resources for Sustainability
Mars has abundant resources that could support human colonization. Its surface contains water ice, which can be used for drinking, agriculture, and even producing rocket fuel (via the electrolysis of water to create hydrogen and oxygen). The Martian atmosphere, though thin and composed mostly of carbon dioxide, can be harnessed for oxygen production and methane synthesis. These resources reduce the need to transport everything from Earth, making long-term habitation more feasible.

Survival of Humanity
Elon Musk and other advocates of Mars colonization emphasize the existential imperative of becoming a multi-planetary species. Earth faces numerous risks, including climate change, asteroid impacts, pandemics, and geopolitical conflicts. Establishing a self-sustaining colony on Mars provides a "backup plan" for humanity, ensuring our survival even in the face of catastrophic events.

The Technical Challenges of Reaching Mars

Sending humans to Mars is an engineering challenge of unprecedented scale. From propulsion systems to spacecraft durability, every aspect of the mission requires careful planning and innovation.

Interplanetary Propulsion
The journey to Mars takes approximately six to nine months, depending on the relative positions of Earth and Mars. Starship's Raptor engines, which use liquid methane and liquid oxygen (CH_4/LOX), are optimized for interplanetary travel. This choice of fuel is critical, as methane can potentially be produced on Mars, enabling return trips and reducing the reliance on Earth-based resources.

The spacecraft must also be capable of carrying sufficient cargo and crew for the journey. Starship's payload capacity—over 100 tons to Mars—makes it uniquely suited for this mission. Its spacious interior can accommodate life support systems, supplies, and habitat modules, providing the infrastructure needed for long-duration missions.

Radiation Protection
One of the most significant hazards of interplanetary travel is exposure to cosmic radiation and solar particle events. Unlike Earth, which is shielded by a robust magnetic field and thick atmosphere, spacecraft traveling to Mars are vulnerable to harmful radiation. Starship's design includes protective shielding and the use of water or other materials to create safe zones for the crew during periods of high radiation.

Orbital Mechanics and Landing
Landing on Mars presents unique challenges due to its thin atmosphere, which is too sparse for effective aerodynamic braking but still thick enough to generate heat during descent. Starship's aerodynamic design, coupled with its use of retropropulsion for landing, addresses these challenges. The spacecraft's "belly-flop" maneuver, tested extensively during Earth-based prototypes, will be critical for ensuring a safe and controlled descent.

The Logistical Challenges of Colonizing Mars

Establishing a permanent human presence on Mars requires solving logistical challenges on a scale never before attempted. From transporting materials to building infrastructure, every step must be carefully planned to ensure sustainability.

Transporting Materials and People
Starship's reusability and high payload capacity are central to SpaceX's plan for Mars colonization. By reducing the cost of launches, SpaceX aims to make repeated trips to Mars economically viable. Initial missions will focus on transporting essential equipment, such as power generators, habitats, and construction machinery, to establish the foundation for a colony.

SpaceX envisions fleets of Starships traveling together during favorable launch windows, which occur approximately every 26 months. These windows, dictated by the alignment of Earth and Mars, allow for the most efficient transfer between the planets. Coordinating these missions will require meticulous planning and advanced navigation systems.

Energy and Power Generation
Sustainable energy production is critical for a Martian colony. Solar power is the most viable option, as Mars

receives enough sunlight to support photovoltaic systems. However, dust storms on Mars, which can last for weeks or even months, pose a challenge to solar power generation. To address this, backup systems such as nuclear power plants may be deployed to ensure a continuous energy supply.

Food and Water Production
Transporting food from Earth for a large colony is impractical, making local food production essential. Greenhouses equipped with hydroponic or aeroponic systems could be used to grow crops using water extracted from Martian ice. These systems will need to be highly efficient, recycling nutrients and water to minimize waste.

Access to water is equally critical. Ice deposits, identified by orbiters and rovers, will be the primary source of water for early missions. Advanced drilling and extraction technologies will be required to harvest and purify this water for drinking, agriculture, and industrial use.

Habitat Construction
Building habitats that can withstand Mars' harsh environment is a monumental challenge. The thin atmosphere provides little protection from radiation, and temperatures can drop to -100 degrees Celsius at night. Initial habitats may consist of inflatable modules

covered with Martian regolith to provide insulation and radiation shielding.

In the long term, 3D printing technologies could be used to construct permanent structures from locally sourced materials. These habitats would need to be airtight, temperature-controlled, and equipped with life support systems to maintain a livable environment for the inhabitants.

The Human Challenges of Mars Colonization

While technology can solve many of the logistical and engineering challenges, the human element of Mars colonization presents its own set of difficulties. The psychological and physiological impacts of living on another planet cannot be overlooked.

Isolation and Mental Health

Mars is, at best, a seven-month journey from Earth, and communication delays range from 4 to 24 minutes each way. These factors create a profound sense of isolation for colonists, who must adapt to living in a small community with limited interaction with friends and family on Earth.

Maintaining mental health in such an environment will require careful planning. Colonists will need access to entertainment, social activities, and mental health resources. Establishing a sense of purpose and community will be critical for fostering resilience and well-being.

Adaptation to Low Gravity
Mars' surface gravity is about 38% of Earth's, which could have long-term effects on human health. Prolonged exposure to low gravity can lead to muscle atrophy, bone density loss, and cardiovascular changes. Countermeasures, such as exercise regimens and artificial gravity habitats, may be necessary to mitigate these effects.

Cultural and Ethical Considerations
Mars colonization will inevitably raise questions about governance, culture, and ethics. How will the first Martian society be organized? What rights and responsibilities will colonists have? How do we ensure that Mars colonization benefits all of humanity rather than a select few? Addressing these questions will require international cooperation and a commitment to inclusivity and fairness.

Starship: The Cornerstone of Mars Exploration

Starship is more than a spacecraft; it is the foundation of SpaceX's vision for Mars. Its design, capabilities, and versatility make it uniquely suited for the challenges of interplanetary travel and colonization.

Versatility and Scalability
Starship's modular design allows it to serve multiple roles, from cargo transport to crewed missions. Its spacious interior can be customized to carry scientific instruments, life support systems, and habitat modules. This adaptability makes it a critical tool for building the infrastructure needed for a Mars colony.

Cost Reduction Through Reusability
The reusability of Starship is key to making Mars colonization economically viable. By reducing the cost per launch, SpaceX can facilitate the frequent and large-scale transportation of people and materials. This cost-efficiency is essential for sustaining the long-term development of a Martian colony.

In-Situ Resource Utilization
Starship's methane-powered engines enable the use of in-situ resource utilization (ISRU) on Mars. By producing methane and oxygen from Martian water and carbon dioxide, Starship can refuel on the planet, eliminating the need to transport return fuel from Earth. This capability is critical for enabling round-trip

missions and reducing reliance on Earth-based resources.

The Path Forward: Preparing for Humanity's Most Ambitious Journey

SpaceX's plan for Mars is still in its early stages, but each test flight and mission brings us closer to realizing the dream of interplanetary colonization. Key milestones, such as the first crewed missions to Mars and the establishment of initial habitats, will lay the groundwork for a self-sustaining colony.

While the challenges are immense, the potential rewards are equally profound. Colonizing Mars offers humanity a chance to expand its horizons, secure its future, and embark on a journey of exploration and discovery that will define the next chapter of our history.

Mars represents humanity's greatest challenge and its greatest opportunity. Preparing for this journey requires solving technical, logistical, and human challenges on an unprecedented scale. With Starship as its cornerstone, SpaceX is leading the charge, turning the dream of Mars colonization into a tangible goal.

As humanity takes its first steps toward becoming a multi-planetary species, we are not just reaching for another world—we are redefining what it means to explore, innovate, and thrive. The journey to Mars is not just about technology; it is about the enduring spirit of curiosity and resilience that drives us to venture into the unknown.

Chapter 14

Beyond Mars

While Mars represents humanity's most immediate goal for interplanetary exploration, it is only the beginning of a grander journey. Space exploration has always been about more than reaching new destinations; it is about redefining what humanity is capable of, imagining what lies beyond, and charting our place in the cosmos. Beyond Mars, the possibilities stretch into the vast unknown: asteroid mining, establishing outposts on distant moons, building interstellar spacecraft, and perhaps even encountering other intelligent life.

This chapter speculates on humanity's long-term vision for space exploration, examining how innovations like SpaceX's Starship could enable extraordinary ventures. From economic opportunities such as asteroid mining to existential questions about humanity's role in the universe, the future of space exploration promises to be as transformative as it is ambitious.

Asteroid Mining: The Next Frontier of Resource Extraction

Asteroids, those seemingly barren rocks floating in the void, hold immense economic and scientific potential. These celestial bodies are rich in valuable resources, including water, metals, and rare-earth elements. Unlocking the wealth of asteroids could revolutionize industries on Earth and provide the raw materials needed for large-scale space exploration.

The Economic Potential of Asteroids

Asteroids are treasure troves of resources. For example, some metallic asteroids contain vast quantities of platinum, gold, and other precious metals, with individual asteroids estimated to hold trillions of dollars' worth of materials. Beyond metals, asteroids contain water ice, which can be broken down into hydrogen and oxygen to create rocket fuel. This could enable the development of "space gas stations," reducing the need to carry all fuel from Earth and making long-distance space travel more feasible.

The economic implications of asteroid mining are staggering. By providing access to materials that are scarce or expensive to extract on Earth, asteroid mining could reduce resource scarcity, stabilize markets, and drive technological innovation. Companies like Planetary Resources and Deep Space Industries are already exploring this potential, with SpaceX's Starship envisioned as a critical tool for transporting mining equipment and materials.

Challenges and Innovations

While the promise of asteroid mining is immense, the challenges are equally daunting. Mining operations in microgravity require entirely new technologies for excavation, processing, and transport. Moreover, identifying viable asteroids and developing the infrastructure to extract resources will require significant investment and collaboration.

Starship could play a pivotal role in overcoming these challenges. Its ability to carry large payloads and deploy autonomous mining equipment to asteroids positions it as an ideal vehicle for pioneering asteroid mining missions. Over time, these missions could form the backbone of a space-based economy, enabling sustainable exploration and colonization.

Ethical and Legal Considerations

Asteroid mining also raises ethical and legal questions. Who owns the resources extracted from space? How do we ensure that the benefits of asteroid mining are distributed equitably? Existing treaties, such as the Outer Space Treaty, do not provide clear answers, and new frameworks will be needed to regulate the exploitation of space resources.

Colonizing the Outer Solar System

While Mars offers the most immediate opportunity for colonization, the outer Solar System holds untapped potential for human settlement. The icy moons of Jupiter and Saturn—such as Europa, Ganymede, Titan, and Enceladus—are among the most intriguing targets for exploration and colonization.

Europa and the Search for Life

Europa, one of Jupiter's largest moons, is a prime candidate for the search for extraterrestrial life. Beneath its icy crust lies a subsurface ocean, potentially harboring conditions suitable for microbial life. Colonizing Europa would require advanced technologies to drill through its ice and establish habitats beneath the surface, where radiation levels are lower.

Starship could serve as a transport vehicle for missions to Europa, delivering scientific equipment, autonomous probes, and eventually human crews. A Europa outpost would not only advance our understanding of life's potential beyond Earth but also serve as a stepping stone for exploring the outer Solar System.

Titan: A Second Earth?

Saturn's moon Titan is another compelling target for colonization. With a thick atmosphere and surface lakes of liquid methane and ethane, Titan offers unique opportunities for exploration. While its extreme cold poses challenges, its atmosphere provides protection from radiation, and its abundant hydrocarbons could serve as a valuable resource for energy production.

Establishing a colony on Titan would be an ambitious endeavor, requiring self-sustaining habitats, advanced energy systems, and efficient transport solutions. Starship's capacity for long-duration missions and heavy payloads makes it a strong candidate for supporting these efforts.

Mining and Fuel Production on Distant Moons
Beyond colonization, the outer moons could serve as hubs for resource extraction and fuel production. The presence of water ice on moons like Europa and Enceladus could support hydrogen-based fuel production, enabling further exploration of the outer Solar System and beyond. These moons could become key nodes in a growing network of interplanetary infrastructure.

Interstellar Travel: Humanity's Ultimate Challenge

While our Solar System offers countless opportunities for exploration and settlement, the ultimate frontier lies beyond: the stars. Interstellar travel represents humanity's most ambitious goal, requiring technological breakthroughs that currently exist only in theory.

The Challenges of Interstellar Travel
The vast distances between stars present an extraordinary challenge. Even the closest star system, Alpha Centauri, is 4.37 light-years away. With current propulsion technologies, such a journey would take tens of thousands of years. Achieving interstellar travel will require new propulsion systems capable of reaching a significant fraction of the speed of light.

Potential solutions include *nuclear fusion propulsion*, *solar sails*, and *antimatter engines*. Each of these concepts faces significant scientific and engineering hurdles, from fuel production to managing the effects of high-speed travel on spacecraft structures.

Starship as a Stepping Stone
While Starship is not designed for interstellar missions, its development lays the groundwork for the technologies and infrastructure needed for such endeavors. The modularity, reusability, and scalability of Starship provide a model for future interstellar spacecraft, demonstrating how cost-efficient exploration can expand humanity's reach.

The Search for Habitable Exoplanets
The discovery of exoplanets in the habitable zones of distant stars has renewed interest in interstellar exploration. Missions like NASA's Kepler and TESS have identified thousands of exoplanets, some of which may possess conditions suitable for life. Establishing an interstellar presence would enable humanity to study these worlds up close and perhaps even establish colonies beyond our Solar System.

Humanity's Role in the Cosmos

Beyond the technical and logistical challenges of exploration lies a deeper question: What is humanity's role in the cosmos? As we venture further into space, we are confronted with profound philosophical and existential questions.

Expanding Human Consciousness
Space exploration is more than a quest for survival—it is a journey of discovery that expands our understanding of the universe and our place within it. By venturing beyond Earth, we gain new perspectives on our origins, our shared humanity, and the possibilities that lie ahead.

Preserving and Spreading Life

Becoming a multi-planetary species is not just about preserving humanity; it is about spreading life itself. As the only known intelligent beings in the universe, we have a unique responsibility to protect and propagate the spark of life. Colonizing other planets and moons, and perhaps even sending life to distant star systems, ensures that life continues to thrive in the face of cosmic challenges.

Ethical and Cultural Considerations
With great ambition comes great responsibility. As we explore and settle new worlds, we must ensure that our actions reflect the best of our values. This includes protecting the environments of other celestial bodies, respecting the potential for alien life, and fostering equitable access to the benefits of space exploration.

The Role of Starship in Humanity's Cosmic Future

Starship is not just a spacecraft—it is a symbol of what humanity can achieve when we dare to dream big. Its versatility, reusability, and cost-efficiency position it as a cornerstone for future exploration, from asteroid mining to interstellar travel.

Building an Interplanetary Network

Starship's ability to establish colonies, transport resources, and support long-duration missions makes it a key player in building an interplanetary network. This network could include refueling stations, research outposts, and industrial hubs, enabling seamless travel and commerce across the Solar System.

Inspiring the Next Generation

Starship has already inspired millions to think beyond Earth and imagine humanity's future in space. By achieving milestones like Mars colonization and resource extraction, Starship will continue to ignite curiosity and ambition, encouraging future generations to pursue careers in science, engineering, and exploration.

A Journey Without End

The journey beyond Mars is not a destination—it is an ongoing process of exploration, discovery, and growth. From mining asteroids to colonizing distant moons, from interstellar travel to the search for life beyond Earth, the possibilities are as vast as the universe itself.

As humanity ventures into the cosmos, we carry with us not just our technology but also our values, aspirations, and dreams. Starship, as a tool and a symbol, represents the beginning of this journey, offering a glimpse of what lies ahead. The future of humanity is not bound by the limits of Earth—it is written in the stars.

Conclusion

The Starship Legacy

As the Starship program evolves, its significance extends far beyond the bounds of space exploration. Starship represents a confluence of ambition, ingenuity, and determination that challenges humanity to rethink what is possible—not only in aerospace but across every industry. By redefining the limits of innovation, SpaceX has established a legacy that transcends its technical achievements, serving as a testament to human resilience and the power of bold vision.

The journey of Starship is not just the story of a spacecraft—it is the story of a paradigm shift in how we approach challenges, solve problems, and work toward a shared future. In reflecting on the broader impact of Starship's development, we find lessons that resonate far beyond the launchpad. These lessons offer a roadmap for future innovations, while the program itself serves as a powerful reminder of what humanity can achieve through curiosity, ambition, and collaboration.

A Transformative Vision

From the beginning, Starship has been a project that defies convention. Its scale, ambition, and purpose set it apart from any previous spacecraft. Starship is not merely a vehicle to reach Mars—it is a tool to reshape humanity's relationship with the cosmos. By striving to make life multi-planetary, SpaceX has given the world a unifying goal: the survival and expansion of human civilization.

Redefining Space Exploration
Starship's development has redefined the goals and possibilities of space exploration. Traditional space programs, constrained by budgets, political considerations, and incremental progress, often viewed space exploration as a series of isolated missions. Starship changes this narrative by envisioning a continuous, scalable presence in space, with reusable technology and a focus on sustainability. This shift challenges other organizations, both public and private, to adopt similarly bold visions, sparking a new era of exploration.

Inspiring Global Collaboration
The challenges posed by space exploration—whether technical, logistical, or philosophical—are too vast for any one nation or organization to tackle alone. Starship's mission underscores the importance of global

collaboration. While SpaceX is a private company, its achievements have catalyzed international interest and dialogue, encouraging partnerships that transcend borders. The Artemis program, which involves collaboration between NASA, ESA, and other international agencies, exemplifies the potential for cooperative efforts driven by shared goals.

Lessons from Starship: Innovation and Resilience

The development of Starship offers profound lessons that extend far beyond the aerospace industry. SpaceX's approach to innovation, risk-taking, and resilience provides a template for tackling complex challenges in any field.

The Power of Iteration

At the heart of Starship's success is SpaceX's philosophy of rapid prototyping and iterative design. Each prototype, whether it succeeded or failed, contributed valuable data that informed the next version. This approach contrasts sharply with traditional methods, which often prioritize perfection at the expense of speed. By embracing failure as a tool for learning, SpaceX has demonstrated that progress is not a straight line but a process of constant refinement.

This lesson is applicable across industries, from software development to medicine. In a world where challenges are increasingly complex and interconnected, the ability to iterate quickly and adapt to new information is essential for success.

Risk-Taking as a Catalyst for Progress
Starship's development has been marked by bold risks, from testing new materials to attempting ambitious flight maneuvers. These risks were calculated, rooted in rigorous analysis and a willingness to push the boundaries of what was thought possible. By challenging conventional wisdom and taking risks, SpaceX has achieved breakthroughs that might never have been possible under a more cautious approach.

Organizations in other industries can learn from this example, embracing calculated risks as a means of driving innovation. Whether it's developing renewable energy solutions, tackling climate change, or addressing global health crises, bold action is often the catalyst for transformative progress.

Resilience in the Face of Setbacks
Starship's journey has not been without setbacks. From explosive test flights to engineering challenges, each obstacle tested the resilience of SpaceX's team. Yet, each failure was met with determination, analysis, and an unwavering commitment to the mission. This resilience

is a reminder that success is not about avoiding failure but about how we respond to it.

In a rapidly changing world, resilience is a critical trait for individuals, organizations, and societies. By viewing setbacks as opportunities to learn and grow, we can build the strength needed to tackle even the most daunting challenges.

A Broader Impact: Inspiring Future Innovations

The influence of Starship extends far beyond the field of aerospace. Its innovations in materials science, propulsion systems, and manufacturing processes are already inspiring advancements in other industries, from transportation to energy. More importantly, the ethos of curiosity and ambition that drives the program has the potential to reshape how we approach innovation across every field.

Cross-Industry Applications
- Sustainability and Reusability: Starship's focus on reusability has sparked interest in applying similar principles to other industries. For example, the automotive and construction sectors are exploring ways to design products with longer

lifespans and recyclable components, reducing waste and environmental impact.
- Autonomous Systems: The autonomous technologies developed for Starship, such as navigation and landing systems, have applications in robotics, shipping, and urban mobility.
- Energy Innovation: The methane-based propulsion systems used in Starship highlight the potential of alternative fuels. These technologies could pave the way for advancements in renewable energy and clean fuel production on Earth.

A Catalyst for Education and Exploration

Starship has already inspired countless young people to pursue careers in science, technology, engineering, and mathematics (STEM). By capturing the public's imagination, the program has ignited a renewed interest in exploration, discovery, and problem-solving. Educational initiatives, public outreach, and partnerships with schools and universities ensure that Starship's legacy will extend to the next generation of innovators.

The Importance of Global Collaboration

As humanity ventures further into space, the importance of global collaboration cannot be overstated. The challenges we face—both on Earth and beyond—require collective effort and shared resources. Starship's mission highlights the potential for partnerships that transcend borders, combining the strengths of nations, organizations, and individuals to achieve common goals.

Space as a Unifying Force

Space exploration has always had the power to bring people together. The sight of Earth from the Moon, captured during the Apollo missions, inspired a sense of shared humanity that transcended national boundaries. Starship has the potential to evoke a similar sense of unity, reminding us that the challenges we face are not confined to any one nation but shared by all.

Collaboration in the Cosmos

The future of space exploration will depend on international partnerships. From agreements on resource sharing to joint missions to other planets, collaboration will be essential for ensuring that space remains a domain of peace and discovery. Starship, as a cornerstone of future exploration, will play a critical role in fostering these partnerships.

A Call to Action: Curiosity, Ambition, and Unity

As the Starship program moves closer to realizing its goals, it serves as a powerful reminder of the importance of curiosity, ambition, and unity in exploring the unknown. Humanity has always been driven by a desire to understand the world around us, to push boundaries, and to seek out new frontiers. Starship embodies these qualities, offering a glimpse of what we can achieve when we dare to dream big.

Fostering Curiosity
Curiosity is the foundation of exploration. It drives us to ask questions, seek answers, and imagine what lies beyond. By fostering curiosity in ourselves and others, we can unlock new possibilities and inspire the next generation to pursue their own journeys of discovery.

Embracing Ambition
Ambition fuels progress. It pushes us to set audacious goals and work tirelessly to achieve them. Starship is a testament to the power of ambition, showing that even the most daunting challenges can be overcome with vision, determination, and collaboration.

Promoting Unity
In a world often divided by conflict and competition, space exploration offers a rare opportunity to unite around a common goal. By working together, we can ensure that the benefits of exploration are shared by all, creating a future that reflects the best of humanity.

The Starship Legacy

The legacy of Starship is not just about reaching Mars or building colonies—it is about inspiring humanity to think beyond the limits of today and imagine the possibilities of tomorrow. Through its bold vision, groundbreaking technology, and unwavering commitment to progress, SpaceX has shown us what we are capable of when we work together to explore the unknown.

As we look to the future, Starship reminds us that the journey is just beginning. The challenges we face, both on Earth and in the cosmos, are immense. But so too are the opportunities. By embracing curiosity, ambition, and unity, we can continue to push the boundaries of what is possible, ensuring that the legacy of Starship endures for generations to come. The stars are not as far away as they once seemed—and together, we can reach them.

Appendix

The appendix provides additional details, technical insights, and supporting information for readers seeking a deeper understanding of the topics discussed in this book. It serves as a resource for enthusiasts, educators, and professionals interested in the intricacies of space exploration and the Starship program.

Appendix A: Technical Specifications of Starship

- Height: 120 meters (394 feet), including the Super Heavy booster.
- Diameter: 9 meters (30 feet).

Payload Capacity:
- To Low Earth Orbit (LEO): Over 100 metric tons.
- To Mars: Estimated 100 metric tons, depending on mission profile.

Propulsion:
- Engines: 33 Raptor engines on the Super Heavy booster, 6 Raptor engines on Starship (3 optimized for sea level, 3 optimized for vacuum).
- Propellants: Liquid methane (CH_4) and liquid oxygen (LOX).

- Thrust: ~16 million pounds (Super Heavy and Starship combined).
- Reusability: Designed for full reusability, with rapid turnaround between missions.

Primary Applications:
 - Crewed Mars missions.
 - Lunar landings for NASA's Artemis program.
 - Deployment of large satellite constellations (e.g., Starlink).
 - Point-to-point Earth transportation.

Appendix B: Key Milestones in the Starship Program

1. *2012*: Elon Musk announces plans for a fully reusable rocket system to support interplanetary travel.
2. *2016*: SpaceX unveils the Interplanetary Transport System (ITS), the precursor to Starship.
3. *2018*: The first Starship prototype, "Starhopper," begins development.
4. *2019*: Starship Mk1 is revealed; early suborbital tests conducted.
5. *2020*: Starship SN5 and SN6 complete successful low-altitude test flights.
6. *2021*: Starship SN15 achieves the first fully successful high-altitude test and landing.
7. *2023*: First integrated flight test of Starship and the Super Heavy booster.

8. *2024 and Beyond*: Planned Mars cargo missions, lunar landings, and crewed Mars missions.

Appendix C: Glossary of Terms

- Belly-Flop Maneuver: A controlled aerodynamic descent used by Starship to maximize drag and slow down during re-entry.
- ISRU (In-Situ Resource Utilization): The practice of using local resources (e.g., water and carbon dioxide on Mars) to support space missions.
- LEO (Low Earth Orbit): An orbit around Earth at an altitude of approximately 160–2,000 kilometers (100–1,200 miles).
- Methalox: A type of rocket fuel combination consisting of liquid methane and liquid oxygen, used by Starship's Raptor engines.
- Raptor Engine: A full-flow staged combustion rocket engine developed by SpaceX, optimized for reusability and high performance.

Appendix D: SpaceX's Key Contributions to Space Exploration

1. *Reusable Rockets*: Falcon 9 and Falcon Heavy demonstrated the feasibility of reusability, significantly reducing launch costs.

2. *Commercial Crew Program*: SpaceX's Crew Dragon became the first privately built spacecraft to transport astronauts to the ISS.
3. *Starlink*: Deployment of a global satellite network to provide high-speed internet access.
4. *Starship*: Designed to enable interplanetary travel, including missions to Mars and beyond.

Appendix E: Timeline of Space Exploration

- 1957: The Soviet Union launches Sputnik, the first artificial satellite.
- 1961: Yuri Gagarin becomes the first human in space.
- 1969: Apollo 11 lands the first humans on the Moon.
- 1981: NASA launches the Space Shuttle, the first reusable spacecraft.
- 1998: The International Space Station (ISS) begins construction.
- 2010: SpaceX becomes the first private company to return a spacecraft from orbit.
- 2020: SpaceX's Crew Dragon completes its first crewed mission to the ISS.
- Future: Human missions to Mars, asteroid mining, and interstellar exploration.

Appendix F: Recommended Resources

Books
- The Case for Mars by Robert Zubrin
- The Overview Effect by Frank White
- Elon Musk: Tesla, SpaceX, and the Quest for a Fantastic Future by Ashlee Vance

Documentaries
- Mars: Inside SpaceX (National Geographic)
- The Last Man on the Moon (Netflix)
- Apollo 11 (CNN Films)

Websites
- [SpaceX Official Website](https://www.spacex.com)
- [NASA Artemis Program](https://www.nasa.gov/specials/artemis)
- [European Space Agency (ESA)](https://www.esa.int)

Online Communities
- [r/SpaceX](https://www.reddit.com/r/SpaceX) (Reddit)
- [NASASpaceFlight Forum](https://forum.nasaspaceflight.com)

Appendix G: Ethical Considerations in Space Exploration

- Resource Management: Ensuring that space resources are used sustainably and equitably.
- Planetary Protection: Preventing contamination of other worlds and preserving their natural environments.
- Governance: Establishing international laws and treaties to regulate space activities and prevent conflicts.
- Equity: Ensuring that the benefits of space exploration are shared globally, rather than concentrated among a few nations or corporations.

Appendix H: Frequently Asked Questions About Starship

1. How many people can Starship carry to Mars?
   - Starship is designed to carry up to 100 passengers, depending on mission requirements and cargo needs.
2. What makes Starship different from other rockets?
   - Starship is fully reusable, designed for long-duration missions, and capable of carrying large payloads to multiple destinations, including Mars and the Moon.
3. When will the first crewed Mars mission take place?

- SpaceX aims to send the first crewed mission to Mars in the 2030s, following initial cargo missions and infrastructure development.

4. What happens if something goes wrong during a mission?

- Starship is equipped with advanced safety features, including redundant systems and escape protocols for crewed missions.

Appendix I: The Future of Space Exploration

As humanity ventures further into space, the potential applications of programs like Starship extend to:

- Space-Based Solar Power: Harnessing solar energy in space and transmitting it to Earth.
- Space Tourism: Expanding access to space for commercial and recreational purposes.
- Interstellar Exploration: Developing technologies to reach distant star systems and study exoplanets.
- Terraforming: Investigating the feasibility of altering planetary environments to support human life.

By fostering curiosity, ambition, and collaboration, the Starship program is paving the way for humanity's next great adventure. The appendix highlights the resources, knowledge, and inspiration that will guide us as we continue to explore the final frontier.

About the Author

William A. Sanders is a seasoned writer and lifelong space exploration enthusiast, with a passion for understanding and sharing the groundbreaking advancements that define our era. With a deep interest in the intersection of science, technology, and humanity's potential, Sanders has spent years researching and writing about the transformative impact of space exploration.

Growing up inspired by the stories of Apollo astronauts and the breathtaking images of distant galaxies, Sanders developed an insatiable curiosity about the universe and humanity's place within it. This fascination led him to explore not just the technical aspects of space exploration, but also the philosophical and cultural dimensions of venturing into the cosmos.

In his work, Sanders blends meticulous research with engaging storytelling, making complex topics accessible and compelling for a broad audience. His writing delves into the challenges, triumphs, and ethical questions surrounding humanity's efforts to reach new worlds, offering readers both a deep understanding of the subject and a sense of awe for what lies ahead.

Through his books and articles, Sanders seeks to inspire readers to think boldly, dream big, and embrace the spirit of exploration that drives human progress. Whether writing about the engineering marvels of SpaceX's Starship or the philosophical implications of becoming a multi-planetary species, Sanders brings a unique perspective that resonates with space enthusiasts and newcomers alike.

When not writing, William A. Sanders enjoys stargazing, reading science fiction, and engaging with the vibrant community of space exploration advocates. He believes that humanity's future lies in the stars and is dedicated to sharing that vision with the world.

www.ingramcontent.com/pod-product-compliance
Lightning Source LLC
Chambersburg PA
CBHW071025240526

45469CB00006BD/2100